203 Current Topics in Microbiology and Immunology

Editors

Springer
Berlin
Heidelberg
New York
Barcelona
Budapest
Hong Kong
London
Milan
Paris
Tokyo

Cap-Independent Translation

Edited by P. Sarnow

With 31 Figures

Springer

Peter Sarnow
University of Colorado
Health Sciences Center
Department of Biochemistry,
Biophysics, and Genetics
4200 East Ninth Avenue, Box 172
Denver, CO 80262
USA

Cover illustration: The cover illustration is reproduced from a water-color drawing by Malcolm DuBois. The RNA at the top is translated by recruitment of 40S ribosomal subunits in the absence of a functional cap binding protein (CBP) complex. The RNA at the bottom is translated by direct recruitment of 40S subunits to an internal ribosome entry site. Shown are the three components of the CBP complex (top left) and 40S ribosomal subunits carrying initiation factors and initiator tRNAs.

Cover design: Künkel+Lopka, Ilvesheim

ISSN 0070-217X
ISBN-13 978-3-642-79665-4 e-ISBN-13 978-3-642-79663-0
DOI 10 1007/978-3-642-79663-0

Softcover reprint of the hardcover 1st edition 1995

Library of Congress Catalog Card Number 15-12910

Typesetting: Thomson Press (India) Ltd, Madras
SPIN: 10490037 27/3020/SPS – 5 4 3 2 1 0 – Printed on acid-free paper

Preface

Whether or not an mRNA is translated is often thought to be a simple function of its steady-state concentration and the absence of inhibitory proteins or RNA structures in the 5' and 3' noncoding regions. The articles presented in this volume show an unexpected flexibility of the eukaryotic translational apparatus in the mechanism of translational initiation and provide new opportunities for regulation.

Most or all mRNA molecules synthesized by RNA polymerase II in eukaryotic cells contain a 5' terminal m^7GpppG dinucleotide, also known as the "cap structure." RNAs carrying a cap structure have been shown to be more resistant to attack by exoribonucleases than their uncapped counterparts. Furthermore, the cap facilitates transport of the RNAs from the nucleus to the cytoplasm. In the cytoplasm, the cap functions as a binding site for the cap binding protein complex eIF-4, which enhances the translation of the RNAs by the eukaryotic translational apparatus. Specifically, it has been postulated that binding of eIF-4 to the 5' terminal cap facilitates the recruitment of ribosomal subunits onto the mRNAs via their free 5' ends (cap-dependent translation). Accordingly, uncapped RNAs are generally translated more poorly than capped RNAs. However, during the past 6 years both viral and cellular mRNAs have been discovered that can be translated cap-independently. Such RNAs either lack a 5' terminal cap structure, like picornaviral mRNAs, or contain a cap structure but still require little or none of eIF-4 for efficient translation, like adenoviral and certain cellular mRNAs. It is the topic of this volume to describe the features and properties of mRNAs that can be translated cap-independently and to discuss possible physiological roles of cap-independent translation.

In the first chapter, Jackson and colleagues review the role of the cap structure and cap binding proteins in translational initiation and outline the criteria for cap-independent translation. Specifically, the concepts of cap-independent translation by internal initiation and by low eIF-4 requirement are introduced.

Chapters 2–5 (Hellen and Wimmer, Ehrenfeld and Semler, Belsham and colleagues, and Wang and Siddiqui) describe cap-independent translation by internal ribosome binding of picornaviral and hepatitis C viral mRNAs. Using a proposed unifying nomenclature (WIMMER et al. 1993) for picornaviral type I and type II internal ribosome entry sites (IRESs), these chapters provide a detailed analysis of the structural features of the IRES elements and describe the surprising findings that the polypyrimidine tract binding protein (PTB) and the La antigen, both proteins that are normally involved in nuclear RNA biogenesis, stimulate picornaviral translation in the cytoplasm. Late adenoviral mRNAs are capped, lack extensive secondary structures and can be translated without significant amounts of eIF-4. Schneider describes in chapter 6 the nonlinear scanning of late adenoviral 5' noncoding region, whereby major RNA hairpin structures are skipped by the scanning ribosomal subunits. It has been known for a long time that heat shock treatment of eukaryotic cells results in the inhibition of translation. However, mRNAs encoding the heat shock proteins are selectively translated under this condition. Rhoads and Lamphear describe in chapter 7 the known molecular bases for this phenomenon. A diminished amount of eIF-4 is at least one factor that contributes to cap-independent translation of heat shock mRNAs in heat-shocked cells. Lastly, Iizuka and colleagues describe in chapter 8 the known cellular mRNAs that can be translated cap-independently by internal ribosome binding. That certain yeast mRNAs can be translated by internal initiation in a cell-free system from yeast raises the possibility that internal initiation may function in intact yeast cells.

The examples of mRNAs that can be translated cap-independently indicate that regulation at the translational initiation step can be exerted by affecting the point of ribosomal entry or the mode of ribosomal scanning on an mRNA molecule. The use of cap-independent translation can clearly be influenced by the concentration of cytoplasmic translation factors, such as eIF-4, or of nuclear proteins, such as PTB and La. This raises the question whether mRNAs with long 5' noncoding regions, found in mRNAs that encode many regulatory proteins, are always translated as poorly as is predicted by the 5' cap-dependent initiation model. Can some of these mRNAs be translated more efficiently when the concentration of functional translation factors or other proteins is altered in the cell? Elucidation of the mechanism and biological role of cap-independent translation in eukaryotic cells will reveal the importance of this novel mechanism of gene expression in cell physiology and organismal development.

Denver P. SARNOW

References

Wimmer E, Hellen CUT, Cao X (1993) Genetics of poliovirus. Annu Rev Genet 27: 353–436.

List of Contents

List of Contributors

(Their addresses can be found at the beginning of their respective chapters.)

Cap-Dependent and Cap-Independent Translation: Operational Distinctions and Mechanistic Interpretations

R.J. JACKSON, S.L. HUNT, J.E. REYNOLDS, and A. KAMINSKI

1 Introduction

While most of the other chapters in this book are each concerned with particular systems, or particular mRNA species, the purpose of this chapter is to present a more general and comparative overview, not only with respect to aspects which are currently being well established in the field, but also regarding some of the more immediate questions that need to be addressed in the near future. In reflecting on the subject in the course of writing this chapter, we have become increasingly convinced that "cap-independent translation/initiation" is one of the more unfortunate terms to have passed into general usage in the field. In the first place there is more often than not very considerable ambiguity as to whether the

Department of Biochemistry, Tennis Court Road, Cambrige, CB2 1QW, UK term is being used as an operational definition or as a mechanistic explanation. Moreover when used as a mechanistic description there is very often considerable ambiguity as to whether what is implied is true internal initiation or some form of 5' end dependent scanning mechanism that is somehow independent of a 5'cap structure. Even as an operational definition it is not without problems as it is subject to qualifications according to the stringency with which such criteria are applied. There seem to be mRNAs whose translation is cap-independent according to some criteria but not according to others, and the translation of some of these mRNAs (of which the best studied examples are listed in Table 1) might justifiably be classified, operationally, as "marginally cap-dependent" or "pseudo-cap-independent."

The only valid operational criterion for internal initiation is the dicistronic mRNA assay, and by this assay it should be possible to identify a defined segment of the RNA necessary for internal initiation, the so-called IRES (JANG and WIMMER 1990). (Although, as originally defined, IRES was an acronym for "internal ribosome entry site," we have subverted it to "internal ribosome entry segment," since it is invariably understood as meaning the whole segment required for internal initiation and not just the actual precise site at which the internally initiating ribosome binds). Those mRNAs which fail this test must be presumed to be translated by a mechanism related to 5'end-dependent scanning, and there are reasonably good operational criteria available which can confirm this, yet they are all too rarely used. The best is probably to introduce a new AUG codon, preferably not too proximal to the 5' cap but near it (at ~20 nt), preferably located in a good local sequence context according to the rules established by KOZAK (1989) and preferably introduced by the minimum number of mutations (to minimise potential alterations in secondary structure). If this new AUG codon is in-frame with the main reading frame we would expect the mutation to result in the synthesis of a longer product (predominantly but not necessarily exclusively) with little change in total product yield if scanning were operative. If the new AUG is out-of-frame, then the mutation should very significantly reduce the yield of product from the authentic initiation codon. It is curious that of the pseudo-cap-independent mRNAs listed in Table 1, these fairly decisive tests seem to have been applied only in the case of mRNAs with the adenovirus tripartite leader, which not only fail in the dicistronic mRNA test but score positive in the scanning test and which must therefore be translated essentially by a 5' end-dependent scanning mechanism, as discussed in detail elsewhere in this volume (Schneider, this volume).

2 5' Caps and the Role of eIF-4(F)

In order to evaluate the operational criteria for cap-dependent mRNA translation and to try to understand the mechanistic basis of this mode of translation, it is necessary to consider the role of eIF-4 (formerly known as eIF-4F). As usually

Table 1. Classification of eukaryotic mRNA according to the mode of translation initiation

Cap-dependent translation initiation		
Most cellular capped viral mRNAs		
"Marginally cap-dependent"/"pseudo cap-independent" translation initiation		
Alfalfa mosaic virus RNA 4		AMV RNA 4
Adenovirus late mRNAs (or mRNAs with adenovirus tripartite leader)		
Heat shock protein mRNAs		HSP mRNAs
Satellite tobacco necrosis virus RNA		STNV RNA
Potyvirus RNAs		
Internal initiation of translation (IRES-dependent translation initiation)		
Picornavirus RNAs	Human rhinoviruses	HRV RNA
	Enteroviruses, e.g., poliovirus	PV RNA
	Cardioviruses, e.g., encephalomyocarditis virus	EMCV RNA
	Aphthoviruses (foot and mouth disease virus)	FMDV RNA
	Hepatitis A virus	
Hepatitis C virus RNA		HCV RNA
Coronavirus 3a/b/c mRNA		
Immunogloublin heavy chain binding protein mRNA		BiP mRNA
Antennapedia mRNA (*Drosophila*)		
(Human LINE, L1Hs mRNA)		
Undecided (controversial)		
Plant comovirus RNAs, e.g. cowpea mosaic virus RNA		CPMV RNA

mRNAs are classified as translated by internal initiation if they have been shown to give a positive result in the dicistronic mRNA test. mRNAs whose translation shows little or no dependence on the 5' cap but which have not yet been shown to give a positive result in the dicistronic mRNA assay are classified as "marginally cap-dependent" or "pseudo cap-independent." Although the downstream cistron of the naturally dicistronic L1Hs mRNA appears to be translated by internal initiation, the efficiency is very low (McMillan and Singer 1993). The classification of CPMV M RNA is uncertain, because some results are indicative of internal initiation (Verver et al. 1991, Thomas et al. 1991), whilst others are not (Belsham and Lomonosoff 1991)

isolated, eIF-4 from mammalian cells consists of three polypeptide chains: an eIF-4γ subunit of unknown function, which was formerly named p220 in view of its apparent size of 220 kDa on gel electrophoresis, but is now known to be only 154 kDa from the cDNA sequence (Yan et al. 1992); an eIF-4A subunit; and an eIF-4α subunit (formerly known as eIF-4E), which is the only one to have direct affinity for the 5' cap structure (Rhoads et al. 1993). However, there is some evidence that during the initiation process there may be dissociation of some subunits (Rhoads et al. 1993; Joshi et al. 1994), and thus the three subunit state may only exist at certain stages in the initiation pathway. In the presence of eIF-4B, eIF-4 has bidirectional RNA helicase activities of which the 5'–3' helicase requires a 5' capped substrate and is inhibited by m^7GTP (Rozen et al. 1990). Phosphorylation of eIF-4α on Ser-53 seems to be required for full activity. The Ala-53 mutant, though retaining affinity for m^7GTP-Sepharose, is inactive in all functional assays that have been attempted and appears to be defective in assembly into the eIF-4 complex (Joshi-Barve et al. 1990; Rhoads et al. 1993).

Early observations suggested that capping influenced the efficiency of initiation but did not affect the selection of the 5' proximal AUG triplet as the initiation site. This led to the proposal, in the first version of the scanning ribosome model, that the primed 40S ribosomal subunit with its associated initiation factors and Met-tRNA$_f$ first interacted with the capped 5' end of the mRNA and then undertook linear scanning in a 5'–3' direction until the first AUG triplet was encountered (KOZAK 1978, 1989). Subsequently the appreciation that the 40S ribosomal subunit has no obvious special affinity for the capped 5' end of mRNAs, whilst such affinity is displayed uniquely by initiation factor eIF-4 via its eIF-4α component, has given rise to a subtly modified form of this model. Initiation factor eIF-4 is thought to bind to the capped 5' end of the mRNA and to promote RNA unwinding via the action of its eIF-4A helicase component in concert with initiation factor eIF-4B (SONENBERG 1988). It was previously considered that this unwinding might be continued by free (uncomplexed) eIF-4A, again in concert with eIF-4B (SONENBERG 1988), but recent in vitro translation experiments with dominant negative eIF-4A mutants have thrown some doubt on this (PAUSE et al. 1994). These mutant eIF-4A derivatives were strong inhibitors of translation of all types of mRNA, and the inhibition could be overcome more efficiently by addition of eIF-4 complex than free eIF-4A, leading to the suggestion that eIF-4A recycles through the eIF-4 complex and may only function as a helicase in the translation system when in this complexed state (PAUSE et al. 1994). Irrespective of these details, it is thought that once mRNA unwinding has started in the vicinity of the 5' cap, the primed 40S ribosomal subunit then binds near the 5' end of the mRNA and migrates in a 5'–3' direction either together with the helicases or very closely behind them.

The literature on the influence of 5'caps on mRNA translation efficiency, which goes back almost 20 years, has a few inconsistencies but nevertheless a common theme strongly emerges, namely that there is an interdependency of: (a) the influence of the cap structure on translation efficiency of a given mRNA, (b) the susceptibility of translation of this mRNA to inhibition by cap analogues, (c) the influence of K^+ concentration on translation efficiency of the capped species, and on the decrease in efficiency on decapping or addition of cap analogue, and (d) the requirement for initiation factor eIF-4 for efficient translation (KEMPER and STOLARSKY 1977; CHU and RHOADS 1978; WEBER et al. 1978; BERGMANN and LODISH 1979; HERSON et al. 1979; WODNAR-FILIPOWICZ et al. 1978; SONENBERG et al. 1981; EDERY et al. 1984; SLEAT et al. 1988; FLETCHER et al. 1990; TIMMER et al. 1993b). All these parameters are thought to be related to the role of 5' caps and initiation factor eIF-4 in the unwinding of secondary structure in the 5' NCR.

Efficient translation of a spectrum of different (capped) mRNA species in vitro requires different concentrations of eIF-4 in a way that positively correlates with the putative degree of secondary structure near the 5'end of the mRNA (EDERY et al. 1984; FLETCHER et al. 1990; TIMMER et al. 1993b). Supplementation of reticulocyte lysates with extra eIF-4 increases the translation of α-globin mRNA relative to β-globin mRNA (SARKAR et al. 1984), whilst addition of cap analogues has the converse outcome (WEBER et al. 1978).

Overexpression of eIF-4α in vivo (which is presumed but is not yet proven to result in a parallel increase in eIF-4γ expression and hence an increase in intracellular eIF-4 holoenzyme) results in an increase in the efficiency of translation of mRNAs with stable hairpin loops near the 5' end (KOROMILAS et al. 1992).

The observation that increasing monovalent cation concentration invariably increases the difference between the translation efficiency of the capped versus the uncapped version of the same mRNA and invariably increases the susceptibility to inhibition by cap analogues can be explained by the fact that increasing monovalent cation concentration stabilises RNA secondary structure. A reduction in incubation temperature can also achieve the same outcome (WEBER et al. 1978), although as the equilibrium between open and closed forms of RNA hairpin loops is governed by the relationship $\Delta G_0 = -RT \ln K$ in which the temperature (T) is a Kelvin scale and the realistic limits for in vitro translation on this scale are rather narrow (from ~ 295° to ~ 315° K), it is perhaps not surprising that variations in temperature have a lesser effect than variations in monovalent cation concentration.

It is often supposed that the translation of an uncapped mRNA by a 5' end dependent scanning mechanism would require no involvement of eIF-4(F), and the availability of the uncapped mRNA to a scanning 40S ribosomal subunit would be dependent solely on spontaneous unwinding of the 5' NCR. This would correlate with the general observation that for most mRNAs the optimum K^+ concentration for translation of the uncapped version is lower than for the capped form. However, recent experiments have shown that the addition of eIF-4 does stimulate the translation of uncapped mRNAs in vitro, and in a partially fractionated wheat germ system the translation of uncapped AMV 4 RNA or β-globin mRNA actually required very much higher concentrations of this factor than the capped form (FLETCHER et al. 1990; TIMMER et al. 1993b). One possibility is that the role of eIF-4 in the translation of uncapped RNAs reflects the action of the cap-independent 3'–5' helicase activity of this factor. However, it should be noted that, contrary to what is frequently assumed, in a great many experiments but not all (WODNAR- FILIPOWICZ et al. 1979), the translation of uncapped mRNAs was found to be susceptible to inhibition by cap analogues (LODISH and ROSE 1977; HERSON et al. 1979; BERGMANN and LODISH 1979; FLETCHER et al. 1990). The inhibition is potentiated by increasing salt concentrations, just as in the case of capped mRNAs (KEMPER and STOLARSKY 1977). Thus it seems most likely that eIF-4 performs the same functions for the translation of uncapped mRNAs as for capped, which is presumably the 5'–3' rather than the 3'–5' helicase activity (ROZEN et al. 1990), but as its affinity for the uncapped form is much lower, higher concentrations of eIF-4 are required.

The question of whether eIF-4F has any role in true internal initiation of translation is discussed in a later section.

3 Operational Criteria for Cap-Independent Translation

Several types of assay are widely used for this operational definition, as listed below. All but the first involves studying the effect of perturbation of eIF-4 concentrations or activity on the efficiency of translation of the mRNA under scrutiny.

This first assay is a comparative test of the efficiency of translation of capped versus uncapped forms of the same mRNA species. With the development of transcription vectors for in vitro production of RNA, the assay is applicable to a wide variety of mRNAs. Furthermore, by the use of RNA transfection protocols (GALLIE 1991) or microinjection into Xenopus oocytes (DRUMMOND et al. 1985), it is not limited to in vitro translation assays. The interpretation of the results needs to take into account the possibility that the uncapped form of the RNA may be degraded more rapidly than the capped version, which is a more serious problem in intact cell systems (DRUMMOND et al. 1985; GALLIE 1991) and in certain cell-free systems such as the crude wheat germ extract (FLETCHER et al. 1990). As a means of avoiding this complication, some authors have added a non-methylated cap (GpppG) to the 5' end of the transcripts (LODISH and ROSE 1977; PELLETIER et al. 1988a), which seems to protect against degradation but is inactive in promoting translation initiation. Apart from these complications the test would seem to be a relatively straightforward one of comparing the yield of translation product from the two forms of the mRNA. However, as mentioned above, the difference in translation efficiency caused by the 5' cap structure is dependent on the conditions used, particularly on the concentration of monovalent salts, and to a lesser extent (or to a less well documented extent) on the Mg^{2+} and/or polyamine concentrations (KEMPER and STOLARSKY 1977; CHU and RHOADS 1978; WEBER et al. 1978; BERGMANN and LODISH 1979; WODNAR-FILIPOWICZ et al. 1979). Thus this test is only meaningful if conducted over a range of K^+ concentrations extending to concentrations supra-optimal for translation of the capped species.

The second assay measures the effect of cap analogues (m^7GMP, m^7GDP, m^7GTP or m^7GpppG) on the efficiency of translation of the particular mRNA being tested. It is presumed that these act as inhibitors of eIF-4 function certainly with respect to its 5'–3' helicase activity, but perhaps not the 3'–5' helicase activity (ROZEN et al. 1990). Cap-independent translation is indicated if the addition of analogue causes no inhibition of translation over a reasonable concentration range when due consideration has been taken to compensate for Mg^{2+} chelation. In fact, if anything, addition of cap analogue often results in a stimulation of truly cap-independent translation, i.e. internal initiation (ANTHONY and MERRICK 1991). The probable reason is that nuclease-treated cell-free translation systems contain capped fragments of the endogenous mRNA, which act as competitors for the translation of exogenous mRNA, even though their short size means that no translation product of the fragments is detectable and no significant acid precipitable incorporation of radiolabelled amino acids is measurable. The cap analogue inhibits the translation of these fragments and thus relieves the competition against exogenous cap-independent mRNA translation. As in the previous assay, the susceptibility to cap analogue inhibition can be profoundly influenced by the salt concentration (KEMPER and STOLARSKY 1977; CHU and RHOADS 1978; WEBER et al. 1978; BERGMANN and LODISH 1979; WODNAR-FILIPOWICZ et al. 1979), and claims for cap-independent translation based on such assays are only to be taken seriously if K^+ concentrations at or above the optimum for translation are used (ZAPATA et al. 1991). Under appropriate conditions different mRNA species can be ranked according to the concentration of cap analogue required for 50% inhibition of translation (HERSON et al. 1979; SONENBERG et al. 1981; FLETCHER et al. 1990). This ranking is in general agreement with that established by other criteria, such as the susceptibility of translation to inhibition by high salt concentrations (HERSON et al. 1979) or by antibodies against eIF-4α (SONENBERG et al. 1981), and the comparative efficiency of translation of capped versus uncapped forms of the same mRNA species (FLETCHER et al. 1990).

A third assay examines whether the efficiency of translation of the mRNA is influenced by perturbation of the intracellular eIF-4 concentration. This has been achieved by regulated expression in vivo of an antisense transcript to the 5' untranslated region of eIF-4α mRNA, which results in: (1) a gradual decrease in the cellular content of eIF-4α, reaching close to zero after 48 h; (2) a parallel decrease in the cellular content of p220 (eIF-4γ), and therefore presumably also a decrease in eIF-4(F) complex; and (3) a decrease of overall protein synthesis (DE BENEDETTI et al. 1991; RHOADS et al. 1993). By 48 h, most of the residual protein synthesis was heat shock protein synthesis, and the distribution of heat shock protein (HSP) mRNAs in polysomes showed that the initiation frequency on these mRNAs was higher than in control cells (JOSHI-BARVE et al. 1992; RHOADS et al. 1993). The unavoidable conclusion is that translation of HSP mRNAs requires very little, if any, eIF-4.

The fourth assay determines whether the efficiency of translation of a given mRNA is affected by perturbation of eIF-4 activity, either as a result of cleavage of the eIF-4γ component or dephosphorylation of eIF-4α. Cleavage of the eIF-4γ component is achieved in vivo by infection of the cells with an enterovirus (e.g. poliovirus) or foot-and-mouth disease virus (FMDV) (DEVANEY et al. 1988), or by regulated expression of the poliovirus 2A protease (SUN and BALTIMORE 1989). It has also been achieved in vitro by expression of the FMDV L-protease in the cell-free translation system (SCHEPER et al. 1992; THOMAS et al. 1992) or addition of recombinant protease to the system (LIEBIG et al. 1993). There are several reservations about this approach. In the first place there is controversy whether the viral proteinase (e.g. entero/rhinovirus 2A or FMDV L-protease) cleaves eIF-4γ directly (LIEBIG et al. 1993), or indirectly via a proteinase cascade triggered by 2A (WYCKOFF et al. 1992), although it could be argued that this does not influence the validity of this approach as an operational criterion for cap-independent translation. More serious objections arise from conflicting evidence that cleavage of eIF-4γ in vivo may not necessarily be sufficient for inhibition of translation of capped mRNAs, and that eIF-4 activity in some form may be required for IRES-driven initiation. Both of these problems will be discussed in a later section concerned with the possible role of eIF-4 in internal initiation. Despite these problems, this approach successfully identified BiP mRNA as a cellular mRNA whose translation is by a completely cap-independent internal initiation mechanism (SARNOW 1989; MACEJAK and SARNOW 1991).

Perturbation of eIF-4 activity as a result of dephosphorylation of the eIF-4α component occurs late in adenovirus infection and is thought to be the prime cause of the shut-off of host cell mRNA translation (HUANG and SCHNEIDER 1991;

Schneider, this volume). Again, a complication in the evaluation of these data is that the dephosphorylation does not go to completion (Huang and Schneider 1991), and thus we do not know whether such protein synthesis, as continues at this stage in the infectious cycle, is supported by the low amount of phosphorylated factor or whether the dephosphorylated factor retains some activity towards selected mRNA species. Apart from this there is the added complication that the adenovirus infection may introduce other, as yet uncharacterised, modifications to the translation apparatus. Dephosphorylation of eIF-4α also appears to occur during mitosis (Bonneau and Sonenberg 1987a), but the effect of this on the translation of different mRNA species has not been examined in any detail.

4 Specific Cases of Apparent Cap-Independent Translation

With the above criteria in mind it is worth examining some borderline cases of pseudo-cap-independent translation (Table 1) to see whether the evidence points to true internal initiation or to a 5' end-dependent scanning mechanism of initiation, albeit with a very low requirement for a 5' cap structure.

4.1 Alfalfa Mosaic Virus RNA 4

This RNA (AMV RNA 4) has been studied for a long time as an example of an mRNA translated by what was thought to be a cap-independent mechanism, even though it is naturally capped. It has a fairly short (36 nt) and putatively unstructured 5' NCR (Gehrke et al. 1983), with a high content of U and A residues (56% and 25%, respectively). It is translated in extracts of poliovirus-infected HeLa cells at 40% efficiency relative to the efficiency in control cell extracts (Sonenberg et al. 1982), a far higher relative efficiency than for any other capped mRNA tested, but lower than the relative efficiency of satellite tobacco necrosis virus (STNV) RNA translation (50%) or encephalomyocarditis virus (EMCV) RNA translation (90%). Its translation is unusually resistant to inhibition by high salt concentrations (Herson et al. 1979; Gehrke et al. 1983) and is not only more resistant to inhibition by antibodies to eIF-4 or by cap analogues than any other capped mRNA tested (Herson et al. 1979; Sonenberg et al. 1981; Fletcher et al. 1990), but is comparable to STNV RNA translation in these respects. Nevertheless, translation in a fractionated wheat germ system does show a requirement for eIF-4; the apparent K_m for this factor was lower than that for any other capped mRNA species tested but higher than that for STNV RNA (Fletcher et al. 1990; Timmer et al. 1993a,b). Moreover, in contrast to STNV RNA translation, the apparent K_m for eIF-4 was much higher for the uncapped form than the capped (Fletcher et al. 1990; Timmer et al. 1993a). Taken as a whole the evidence points to a 5' end-dependent scanning mechanism and a distinct requirement for a 5' cap and eIF-4 function, albeit an unusually low requirement.

4.2 Late Adenovirus mRNAs

These mRNAs, or chimaeric mRNAs with the adenovirus tripartite leader, were once considered good candidates for true cap-independent translation, since their translation is efficient in late adenovirus infected cells when the host cell mRNA translation is shut off, as a result of the (not quite complete) dephosphorylation of eIF-4α (HUANG and SCHNEIDER 1991; Schneider, this volume). In addition, superinfection by poliovirus late in the adenovirus infectious cycle fails to suppress the adenovirus infection (CASTRILLO and CARRASCO 1987; DOLPH et al. 1988), implying that the translation of the late adenovirus mRNAs is also resistant to the inhibitory effects of cleavage of eIF-4γ (p220). However, cleavage of eIF-4γ by expression of the FMDV L-protease in vitro results in complete inhibition of translation of chimaeric mRNAs with the adenovirus tripartite leader, but not true internal initiation driven by the EMCV IRES (THOMAS et al. 1992). In addition, according to the dicistronic mRNA assay, albeit with a construct in which the initiation codon of the downstream cistron was not quite in the same position as in the late adenovirus mRNAs, the adenovirus tripartite leader does not support internal initiation in vivo (DOLPH et al. 1990), and the introduction of an AUG codon into the first part of this 5' UTR results in inhibition of initiation at the normal AUG codon, entirely as expected if a scanning ribosome mechanism were operative (Schneider, this volume). Taken as a whole, the evidence points to the adenovirus late mRNAs being translated by a 5' end-dependent scanning mechanism, albeit with either a very low requirement for native eIF-4 or the ability to use eIF-4 dephosphorylated in the α subunit. The possibility that this 5' end-dependent mechanism is actually a nonlinear form of scanning involving a so-called ribosomal shunt (FUTTERER et al. 1993) is discussed elsewhere in this volume (Schneider, this volume).

4.3 Heat Shock Protein mRNAs

These mRNAs have also been considered good candidates for cap-independent translation because: (1) their translation in mammalian cells is more resistant than the majority of cellular mRNAs to the shut-off induced by poliovirus infection (MUÑOZ et al. 1984), though not as resistant as the translation of BiP mRNA (SARNOW 1989); (2) in a yeast cell-free system the efficiency of translation of chimaeric mRNAs with the HSP26 mRNA 5' NCR is less influenced by capping the RNA than is the case with the other mRNAs tested (GERSTEL et al. 1992); (3) their translation persists and is actually enhanced in mammalian cells in which the concentration of eIF-4α has been reduced as a result of regulated expression of an antisense transcript to the eIF-4α mRNA (JOSHI-BARVE et al. 1992; RHOADS et al.

1993). In addition the translation of endogenous HSP mRNAs in Drosophila embryo extracts is highly, but perhaps not absolutely, resistant to inhibition by cap analogues and by antibodies against eIF-4α, in sharp contrast to the behaviour of other mRNAs (ZAPATA et al. 1991). These properties explain the selective translation of HSP mRNAs in heat shocked cells, when the activity and integrity of the eIF-4 complex seems to be compromised, for reasons not yet fully understood (LAMPHEAR and PANNIERS 1991; ZAPATA et al. 1991; Rhoads and Lamphear, this volume). The special features of Drosophila HSP 70 mRNA responsible for efficient translation under heat-shock conditions have been identified as lying entirely within the 247 nt long, A-rich 5' NCR, but deletion analysis suggests that it is a cumulative property of the whole 5' NCR rather than assignable to any particular segments within it (MCGARRY and LINDQUIST 1985; LINDQUIST and PETERSEN 1991). Although a 5' NCR length of 150–250 nt and a 45%–50% A-content is common to all Drosophila HSP mRNAs attempts to mimic its properties with a synthetic A-rich 5' NCR were unsuccessful (LINDQUIST and PETERSEN 1991). Addition of extra sequences at the very 5' end abolished translation in heat-shocked cells if extraneous sequences were added, but not if the extension was achieved by duplication of the 5' proximal sequences (MCGARRY and LINDQUIST 1985; LINDQUIST and PETERSEN 1991). This last result, in particular, suggests, but does not prove, that true internal initiation is probably not operative on these mRNAs, but rather that the proximity of the correct sequences near the cap may be important to maintain an unstructured state at the very 5' end in order to allow initiation by a 5' end-dependent scanning mechanism which requires very little, if any, eIF-4 activity.

4.4 Satellite Tobacco Necrosis Virus RNA

A particularly stringent challenge to our criteria regarding different mechanisms of translation initiation is provided by STNV RNA. It is naturally uncapped and has a 29 nt 5' NCR. Its efficiency of translation in an extract of poliovirus infected HeLa cells was 50% relative to that in a control extract (SONENBERG et al. 1982), a higher relative efficiency than for any capped RNA, but less than for EMCV RNA (90%). The translation of STNV RNA in wheat germ systems is considerably more resistant to inhibition by cap analogues than is the case with most mRNAs (HERSON et al. 1979; FLETCHER et al. 1990), but nevertheless there is some inhibition observed and this is accentuated at high salt concentrations, just as with many capped mRNAs (KEMPER and STOLARSKY 1977). Its translation is, at most, only marginally stimulated by capping the RNA, and in a partially fractionated wheat germ system it has an unusually low apparent K_m for eIF-4, which is unaffected by capping, unlike the case with any other mRNA tested (FLETCHER et al. 1990; TIMMER et al. 1993a). In the wheat germ system, modification to the 5' NCR results in little change in the apparent K_m for eIF-4 yet there is a decrease in translation efficiency which can be largely reversed by capping the RNA (TIMMER et al. 1993a). In contrast, deletions in a ~100nt 3' NCR segment just downstream of the translation termination site result in not only a decrease in translation efficiency but also an increase in the apparent K_m for eIF-4, both effects being reversed by capping the mutated form of the RNA (TIMMER et al. 1993a). Transfer of both the 5' and 3' NCR into a chimaeric construct with a different coding sequence confers a large degree of cap independence of translation in the wheat germ system as well as stimulating translation efficiency (TIMMER et al. 1993a), and a somewhat more modest stimulation of translation efficiency is observed if the STNV 3' NCR motif is transferred into a chimaeric construct with the 5' NCR of tobacco mosaic virus rather than STNV (DANTHINE et al. 1993). Clearly STNV RNA provides perhaps the most demanding challenge to the criteria of whether translation is cap-dependent or not. Nevertheless there is as yet no evidence that STNV RNA is translated by true internal initiation, and until positive results are obtained from dicistronic mRNA assays, the presumption must be that it is translated by a 5' end-dependent mechanism, albeit with no requirement for a 5' cap and only a very low requirement for eIF-4, both properties being conferred by the unusual nature of the 5' NCR and the influence of the 3' NCR motif.

4.5 Potyvirus RNAs

These RNAs have often been considered as possible candidates for translation by an internal ribosome entry mechanism since they resemble the animal picornaviruses in having a genome linked protein at the 5' end rather than a 5'cap. The 144 nt 5' NCR of tobacco etch virus was found to be highly stimulatory to translation, as compared with a synthetic 5' NCR, in both transfected plant protoplasts and rabbit reticulocyte lysates (CARRINGTON and FREED 1990). In the RNA transfection assay, the stimulation was independent of whether an unmethylated Gppp or a normal m^7Gppp cap was added, though consideration might be given to the possibility that methylation of the Gppp cap might have occurred in vivo. Translation in the reticulocyte lysate was in fact slightly stimulated by capping and was somewhat sensitive to inhibition by cap analogues. In the absence of any positive evidence from dicistronic mRNA assays, the most reasonable presumption is that the AU-rich (72%) and G-poor (<10%) 5' NCR is probably unstructured, and allows translation by a 5' end-dependent scanning mechanism which is largely independent of the nature of the 5' end and of eIF-4 function.

5 Can a Given mRNA Be Translated by Both 5' End-Dependent and Internal Initiation Mechanisms?

We need to confront the possibility that a given mRNA may show weak IRES activity in the dicistronic mRNA test and yet give somewhat ambiguous results in any test for translation by 5' end-dependent scanning mechanisms. Could such an mRNA be translated by both mechanisms more or less simultaneously? The picornavirus precedents, which suggest that the secondary structure of the IRES is all important for its function (see below), argue against the idea that the first ribosome could translate an mRNA by internal initiation and the second by cap-dependent mechanisms, or vice versa, since the RNA unwinding necessary to allow ribosome scanning would probably disrupt the secondary structure required for internal ribosome entry. Indeed, this may be the explanation of why capping transcripts with the poliovirus 5' NCR not only fails to stimulate translation initiation but actually inhibits (HAMBIDGE and SARNOW 1991). In addition, in the case of the naturally dicistronic L1Hs mRNA, in which the second open reading frame (ORF) is thought to be translated by internal initiation involving a rather inefficient IRES that is likely to include part of the upstream ORF, it has been found that the frequency of internal initiation of downstream cistron translation can be increased by partial inhibition of upstream ORF translation (MCMILLAN and SINGER 1993). However, these considerations do not rule out possibilities such as a cellular mRNA that might be translated by a 5' end-dependent scanning mechanism during most of the cell cycle but switch to internal ribosome entry during mitosis, when the activity of eIF-4 is supposely diminished as a result of dephosphorylation of the eIF-4α polypeptide (BONNEAU and SONENBERG 1987a).

6 Operational Criteria for Internal Initiation

It seems to be often generally assumed that if the translation of an mRNA is cap-independent by any of the above criteria, its translation must inevitably be by an internal ribosome entry mechanism. This assumption is unwarranted without supporting evidence from assays of the translation of dicistronic mRNAs. This dicistronic mRNA test can be designed in two alternative ways, and the relative merits of each approach can be found in a recent review (KAMINSKI et al. 1994b). (1) A standard reporter cistron can be inserted in an extreme upstream position of the cDNA comprising the 5' NCR and the coding sequences of the mRNA under scrutiny; the test hinges on whether the insertion of the upstream reporter ORF significantly reduces expression of the protein encoded by what is now the downstream cistron (KAMINSKI et al. 1994b). (2) Alternatively, the 5' NCR sequences of the mRNA under examination are inserted between two standard reporter ORFs, and the critical question is whether this insertion increases the expression of the downstream cistron from a very low level up to a level comparable to that of the upstream cistron.

With the picornaviruses, which have come to represent the paradigm for internal initiation, it has been the tradition to insert only the 5' NCR between two standard reporter cistrons, although in the original constructs with the EMCV 5' NCR, the downstream cistron coded for a fusion protein with the NH_2-terminal five amino acid residues originating from the viral polyprotein (JANG et al. 1988). It has subsequently been shown that these five codons are not strictly necessary for internal initiation; deletion of these residues does not necessarily compromise internal initiation unless it brings G-rich sequences in proximity to the authentic initiation codon (HUNT et al. 1993). Nevertheless the position of the functional initiation codon with respect to the upstream IRES sequences is critical for the function of the cardiovirus IRESs (HUNT et al. 1993; KAMINSKI et al. 1990, 1994a). In the case of the enterovirus IRES the position of the inserted IRES element relative to the initiation codon of the downstream cistron is not critical (provided that there are no AUG triplets between the putative ribosome entry site at the 3' end of the IRES and the initiation codon of the downstream ORF), and indeed there is a variable and redundant segment in this position in the actual viral genomes (AGOL 1991).

With the wisdom of hindsight, it is perhaps unfortunate that picornaviruses (especially polioviruses) have become the paradigm for internal initiation, as it seems to have led to some unwarranted assumptions in the design of constructs to test for IRESs in other RNAs. The flexibility allowed in the position of the poliovirus IRES relative to the initiation codon is not a universal characteristic not even of all picornavirus IRESs, as is amply demonstrated by the cardioviruses. A failure to appreciate this point in the design of the dicistronic test constructs is almost certainly the explanation of why one report claims that the non-picornavirus hepatitis C virus lacks an IRES, whilst two others show convincing evidence for an efficient IRES (TSUKIYAMA-KOHARA et al. 1992; YOO et al. 1992; WANG et al. 1993). Even the cardiovirus model may lead to the unwarranted assumption that viral coding sequences are invariably unimportant (HUNT et al. 1993), whereas recent evidence from this laboratory shows that coding sequences can play a very important role in the case of the hepatitis C virus IRES (REYNOLDS J.E., KAMINSKI A., GRACE, K., CLARKE B.E., ROWLANDS D.J. and JACKSON R.J., submitted). How many potential IRESs in cellular mRNAs have been overlooked because only the 5' NCR of the mRNA under scrutiny was placed into the dicistronic test construct and no consideration was given to the possibility that coding sequences might be required?

For a credible claim of internal initiation the insertion of the putative IRES element should increase expression of the downstream cistron to a level comparable to that of the upstream cistron. Cases where the insertion gives a yield of downstream cistron products less than 10% of that of upstream cistron products should be regarded with some suspicion, particularly if an in vitro assay was used. It is known that the reticulocyte lysate system will initiate translation at a number of incorrect sites on some RNAs, of which poliovirus, rhinovirus and hepatitis A virus are perhaps the most notorious examples (DORNER et al. 1984; PHILLIPS and EMMERT 1986; JIA et al. 1991; BORMAN and JACKSON 1993). These aberrant initiation events are especially predominant at high RNA concentrations (PHILLIPS and EMMERT 1986; SVITKIN et al. 1994), or at low salt (JACKSON 1991), and they are much less frequent in cell-free systems from other mammalian cell types. Thus any such example of apparently weak IRES activity needs checking that it is not a case of "capricious" or "illegitimate" internal initiation of this type.

Another potential artefact of the dicistronic mRNA test is the possibility of nuclease cleavage within the intercistronic spacer creating two monocistronic mRNAs rather than the desired dicistronic. Whenever this has been checked, for example by northern blotting, it has never been found to be a serious problem, and in the case of the picornavirus IRESs it can be ruled out by the fact that the numerous upstream AUG triplets in such IRESs would serve to prevent ribosomes from reaching the authentic initiation site on monocistronic RNAs generated by nuclease cleavage. Another potential artefact unique to in vivo expression systems is the possibility that the IRES element in the cDNA includes cryptic promoters which could likewise result in the production of two (capped) monocistronic mRNAs.

False positives could potentially arise if the inserted sequences, rather than promoting internal initiation, were permissive to the resumption of scanning by ribosomes that had translated the upstream cistron. In reality, the fact that the use of uncapped mRNAs with a hairpin loop inserted very close to the 5' end reduces the translation of the upstream cistron to virtually zero whilst having no effect on the yield of downstream cistron product shows that this has not been a problem with the IRESs studied so far (M.T. Howell and R.J. Jackson, unpublished observations).

Although the dicistronic mRNA assay system has proved to be a very successful and accurate test for internal ribosome entry, nevertheless, as our colleague Tim Hunt used to remind us, such constructs do still have a physical 5' end and thus we cannot formally eliminate the possibility that the ribosome first binds to the 5' end of the RNA and then makes a long distance "jump" to the initiation codon of the downstream cistron. The only argument against this notion is the fact mentioned above, that downstream cistron translation efficiency is independent of perturbations at the physical 5' end of the RNA, such as de-capping or the insertion of hairpin loops. Nevertheless, given the improved technology for joining RNA molecules in vitro, thus permitting circularisation (Moore and Sharp 1992), it might be worth testing some well established IRES elements in circular form as the final definitive test that internal initiation is really as we imagine it, quite independent of any physical end of the RNA.

7 General Models for Internal Initiation on the Picornavirus Internal Ribosome Entry Segments

Detailed discussion of each of the two main types of picornavirus IRES (the entero-/rhinovirus IRES and the cardio/aphthovirus IRES) are given elsewhere in this volume (Hellen and Wimmer, this volume; Ehrenfeld and Semler, this volume). The purpose of this section is to emphasise the common features of internal initiation of picornavirus RNA translation. A more detailed discussion can be found in Kaminski et al. (1994b) and Jackson et al. (1994).

All picornavirus IRESs appear to be about 450 nt long, and a general model of how this segment directs internal ribosome entry is given in Fig. 1. The essence of

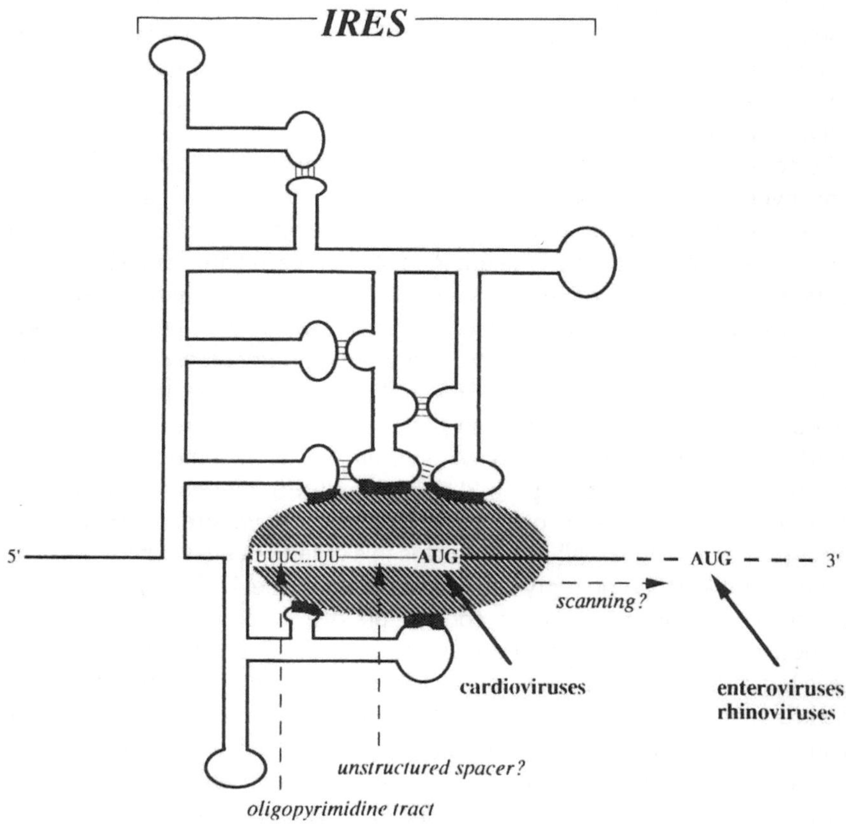

Fig. 1. Internal initiation dependent on the picornavirus internal ribosome entry segments (IRESs). The ~450 nt IRES is viewed as having several base-paired segments, with possible tertiary structure base pairing between these segments. This structure presents a number of short unpaired primary sequence motifs, indicated by *thickened lines*, in the correct spatial organisation. These critical primary sequence motifs may be binding sites for proteins which promote internal initiation, or they may be recognised directly by the ribosome (denoted by the *shaded oval*). The ribosome entry site is at an AUG triplet at the 3' end of the IRES and is located downstream of the oligopyrimidine tract; the segment between this tract and the AUG triplet is thought to serve as an unstructured spacer of defined length. The AUG codon is the authentic initiation site for viral polyprotein synthesis in the case of the cardioviruses. In the entero/rhinoviruses it serves only as a ribosome entry site and is not used as an initiation site; the ribosomes are transferred from this entry site, probably by a scanning mechanism, to the next AUG codon downstream, which is the authentic initiation codon for viral polyprotein synthesis

this model is that most of the ~ 450 nt segment is required by virtue of its secondary, and presumably tertiary, structure in order to present a number of quite short sequence motifs, mostly of them in unpaired regions, in the correct three-dimensional spatial organisation in order to provide a ribosome entry site. This idea is based on the phylogenetic comparisons within each of the two main groups of picornavirus IRESs (the entero/rhinoviruses and cardio/aphthoviruses) which show strong conservation of secondary structure but comparatively few absolutely conserved nucleotide residues, the majority of which are clustered in unpaired regions (Kaminski et al. 1994b; Jackson et al. 1994). The importance of secondary structure is supported by the fact that phenotypic revertants of linker substitution and point mutations often include second-site suppressor mutations which restore the base pairing found in the wild type (Kuge and Nomoto 1987; Haller and Semler 1992).

The actual ribosome entry site is at an AUG codon at the 3' end of the IRES, but the events following ribosome entry at this site differ according to the particular species of picornavirus: (a) in poliovirus and other enteroviruses as well as rhinoviruses, virtually none of the entering ribosomes initiate translation at this AUG triplet, but the ribosomes are transferred, most probably by a scanning mechanism, to the correct initiation codon which is invariably the next AUG codon downstream; (b) in FMDV, about 30% of the entering ribosomes initiate translation at the AUG at the 3' end of the IRES, and the others initiate at the next AUG codon downstream, again most probably after scanning from the entry site AUG triplet (Belsham 1992); (c) in the cardioviruses virtually all the entering ribosomes initiate translation at the AUG at the 3' end of the IRES, though a few may use the next AUG codon located only a short distance downstream (Hunt et al. 1993; Kaminski et al. 1994a). What determines this different behaviour towards the AUG triplet at the ribosome entry site? The context of the AUG may exert some effect, similar to the influence of context on AUG recognition by scanning ribosomes, since the context of the AUG triplet at the 3' end of the cardiovirus IRESs is better than that in polioviruses, according to the precedents established for the scanning ribosome mechanism (Kozak 1989). However, the context of the AUG in poliovirus IRESs is not so poor that we would expect no initiation whatsoever at that site, and indeed, if the relevant segment of the poliovirus 5' NCR is taken out of the IRES and placed into a construct that would be translated by a scanning mechanism, then this AUG is used quite efficiently as an initiation codon (M.T. Howell and R.J. Jackson, unpublished observations). It is almost as if a poor context has a much more serious negative influence on the utilisation of an AUG as an initiation site in the background of an IRES than in a scanning ribosome mechanism, which is reminiscent of the way in which AUG codons located very near the 5'-cap are recognised by scanning ribosomes less efficiently than would be expected on the basis of context criteria alone (Kozak 1991).

In all picornavirus IRESs without exception, the AUG at the putative ribosome entry site is located some 25 nt downstream of the start of a conserved pyrimidine tract, and between this tract and the AUG there is a segment which is quite variable between closely related strains and species but is always G-poor and thus possibly serves as an unstructured spacer of defined length (Pilipenko et al. 1992; Kaminski et al. 1994a,b; Jackson et al. 1990, 1994). Changing the length of this spacer in the EMCV IRES resulted in changes in the choice of initiation site between four closely spaced AUG codons in a way that suggested that the selection was partly, but not wholly, determined by the distance between the pyrimidine-rich tract and the preferred AUG codon (Kaminski et al. 1994a). Similar results have been obtained in the poliovirus system, in which the insertion of extra sequences between the oligopyrimidine tract and the cryptic AUG at the 3' end of the IRES drastically reduced in vitro translation activity and infectivity, but phenotypic revertants arose in which the spacing had been restored, either by deletion or by point mutations within the inserted sequences generating an AUG codon at the appropriate distance (PILIPENKO et al. 1992; GMYL et al. 1993). It is debatable whether the oligopyrimidine tract itself is an important primary sequence determinant of internal ribosome entry or should be regarded as just the start of the unstructured spacer upstream of the AUG triplet. In the case of the poliovirus IRES, a minimal UUUCC motif at the start of the tract has to be retained for efficient in vitro translation and for infectivity (PESTOVA et al. 1991; NICHOLSON et al. 1991; PILIPENKO et al. 1992), but the function of the cardiovirus IRESs is much less affected by complete substitution of the whole tract (KAMINSKI et al. 1994a).

In the case of the poliovirus insertion mutants described above, which reverted by acquiring a new AUG codon as a result of point mutation, it can be inferred that only AUG was acceptable at that site, and neither GUG, UUG or AUU located at the appropriate distance from the pyrimidine-rich tract gave rise to viable viruses (PILIPENKO et al. 1992). However, in the wild-type background, mutation of the AUG to UUG, AAG or AUU reduced the efficiency of translation in vitro from the downstream authentic initiation codon by a maximum of 65%–75% (MEEROVITCH et al. 1991), and mutation to UUG gave a viable virus with a small plaque phenotype (PELLETIER et al. 1988b). Thus other sequences around the cryptic AUG in the wild-type genome must also play a role in the internal ribosome entry mechanism. Interestingly, although UUG and AUU have been shown to function as weak initiation codons in mRNAs translated by cap-dependent mechanisms, AAG has always been found to be inactive in this role (PEABODY 1989). The fact that UUG, AAG or AUU at the putative ribosome entry site of the poliovirus IRES all gave very similar levels of (reduced) translation efficiency suggests that the recognition of the wild-type AUG at this ribosome entry site may involve a different process from the recognition of translation initiation codons by scanning 40S ribosomal subunits.

8 Is eIF-4 Required for Internal Initiation?

A long standing report which predated the discovery of eIF-4 showed that translation of EMCV RNA in a highly fractionated mammalian system required the same set of initiation factors in roughly the same concentrations as did globin mRNA translation (STAEHELIN et al. 1975), and this conclusion is supported by more recent work with partially fractionated systems (SCHEPER et al. 1992). As for eIF-4, the fact that infection by poliovirus or FMDV results in a cleavage of the eIF-4γ component, which roughly correlates with the selective shut-off of host cell mRNA but not viral RNA translation (DEVANEY et al. 1988), leads naturally to the assumption that eIF-4 cleaved in its largest subunit is inactive for cap-dependent translation and redundant for IRES-driven translation. The first of these assumptions has been questioned by the observation that in certain regimes of poliovirus infection, virtually complete cleavage of eIF-4γ can occur with only a modest decrease in host cell mRNA translation (Bonneau and Sonenberg 1987b; Perez and Carrasco 1991). If eIF-4 is as limiting to the translational capacity as is generally assumed, this result seems incompatible with the idea that eIF-4γ cleavage inactivates the factor, and it leads to the suggestion that the normal shut-off of capped mRNA translation requires some other virus-induced perturbation of the translation apparatus, in addition to eIF-4γ cleavage. Nevertheless, it has been shown that extracts of poliovirus infected HeLa cells cannot provide active eIF-4 activity to a fractionated cell-free extract translating a capped mRNA (Etchison et al. 1984). Partially purified eIF-4 from infected cells, which consisted of eIF-4α complexed with the eIF-4γ cleavage products, neither stimulated nor inhibited translation of capped mRNAs in reticulocyte lysates, but slightly stimulated poliovirus RNA translation (Buckley and Ehrenfeld 1987). Of more critical relevance is the fact that overexpression of poliovirus 2A in transfected cells, or addition of recombinant picornavirus 2A to cell-free systems, or expression of FMDV L-protease in such systems, all lead to eIF-4γ cleavage, a strong inhibition of capped mRNA translation, and no inhibition (or more usually a stimulation) of IRES-driven translation (Sun and Baltimore 1989; Thomas et al. 1992; Liebig et al. 1993). Thus even if eIF-4γ cleavage is in itself not sufficient to inactivate the factor selectively for cap-dependent initiation as opposed to internal initiation, then the poliovirus 2A or FMDV L-protease alone must be sufficient to supply the missing function to achieve such inactivation.

All the observations listed above are consistent with the other widely held assumption that eIF-4 activity is not required for internal initiation driven by a picornavirus IRES. Nonetheless, it has been reported that addition of eIF-4 to a reticulocyte lysate or partially fractionated system stimulates internal initiation (Anthony and Merrick 1991; Thomas et al. 1991; Scheper et al. 1992). However, as this internal initiation was mostly with IRESs that function inefficiently in the reticulocyte lysate, either synthetic IRESs, the poliovirus IRES, or the so-called IRES in cow pea mosaic virus (CPMV) M RNA, it might be due to the "capricious" or "illegitimate" internal initiation sometimes seen in such lysates as discussed above. Indeed, supplementation of reticulocyte lysates with additional eIF-4 stimulated initiation at the incorrect sites on poliovirus RNA rather than at the authentic site (Svitkin et al. 1994). However, addition of eIF-4 complex or eIF-4α to a partially fractionated reticulocyte system was reported to stimulate internal initiation driven by the EMCV IRES (Scheper et al. 1992), although in another report from the same group the stimulation seemed to be much less, or saturated at lower levels of added eIF-4 (Thomas et al. 1992). A further indication of a possible role for eIF-4 in internal initiation comes from experiments in which dominant negative eIF-4A mutants were added to reticulocyte lysates: these strongly inhibited the translation of all mRNAs, whether cap-dependent or IRES-dependent, and in all cases the inhibition was reversed much more effectively by wild-type eIF-4 than free eIF-4A (Pause et al. 1994). Thus there are growing suspicions that eIF-4 function may be required for internal initiation, but if this is the case, then the fact that the viral 2A or L-proteases selectively inhibit only cap-dependent initiation implies that eIF-4 with a cleaved eIF-4γ component must be able to carry out this function in internal initiation (THOMAS et al. 1992; LIEBIG et al. 1993).

9 *Trans*-Acting Factors Specific for Internal Initiation

The apparent inability of any picornavirus IRES to function in the wheat germ system, which is fully competent for translation initiation by the 5' end-dependent scanning mechanism, implies that internal initiation, at least on these IRES elements, may need *trans*-acting protein factors in addition to the canonical initiation factors that catalyse initiation on most cellular mRNAs. In addition, the low efficiency and low fidelity of translation of enterovirus and rhinovirus RNA translation in unsupplemented rabbit reticulocyte lysate (DORNER et al. 1984; PHILLIPS and EMMERT 1986; BORMAN et al. 1993), which utilises the EMCV and FMDV IRESs very efficiently, suggests that there may be specific *trans*-acting factor requirements for the different subgroups of picornaviruses. However, this distinction between the entero/rhinovirus IRESs and the cardio/aphthovirus IRESs in the reticulocyte lysates may not be as absolute as is sometimes supposed. It needs to be emphasised that poliovirus RNA is translated quite efficiently and accurately at very low RNA concentrations (PHILLIPS and EMMERT 1986; SVITKIN et al. 1994) and, even in the case of EMCV RNA, saturation of the translation capacity is achieved at rather low RNA concentrations in comparison to typical capped mRNAs translated by a scanning mechanism. Thus a case could be made that initiation on the poliovirus and EMCV IRESs requires the same set of *trans*-acting factors which are present in relatively low concentrations in the reticulocyte lysate and which have a much higher affinity for the EMCV IRES than the poliovirus IRES.

Trans-acting factors required for IRES utilisation are best investigated by functional assays. A more common approach is to examine the binding of proteins to the IRES element, using UV cross-linking or gel retardation assays, and then to try to establish the functional significance of such binding by testing mutant IRES elements. Both methods of investigating binding proteins have strong disadvantages. With UV cross-linking methods it is sometimes easy to forget that the method only detects those proteins which bind in a suitable orientation and proximity for formation of a covalent bond by UV irradiation, and many binding proteins may go undetected. Nonetheless, it has the advantage that the full-length IRES can be used as the target RNA, whereas gel retardation assays, though detecting all reasonably high affinity complexes, have the disadvantage that only relatively short subdomains of the whole IRES can be used. Not only will the use of subdomains eliminate the protein/RNA interactions dependent on long range tertiary structure interactions in the RNA, but the results of UV cross-linking assays with subdomains provides a salutory warning that subdomains may bind or cross-link to proteins which are not cross-linkable to the whole IRES. The pattern of proteins cross-linkable to the intact IRES is not the sum of the proteins cross-linkable to all the individual domains but is usually much simpler. This probably reflects the fact that isolated domains adopt a different folding than in the intact IRES, thus allowing the binding and cross-linking of proteins that are not bound or are not cross-linkable to the intact IRES. As a means of getting round this inherent disadvantage of the gel retardation assay, HALLER and SEMLER (1992) showed that unlabelled intact IRES could outcompete the appearance of the gel retardation complex formed with radiolabelled domain VI, as defined in this volume by Ehrenfeld and Semler (this volume), or domain VII according to the numbering systems of Belsham et al. (this volume).

The only way in which to demonstrate that RNA binding proteins detected by these methods are functionally important in internal initiation is to show strict correlation between loss (or gain) of protein binding and loss (gain) of internal initiation in mutated IRESs bearing subtle mutations rather than gross deletions. This goal has only been attained in one case: the binding of pyrimidine tract binding protein (PTB) to the EMCV IRES was shown to be abrogated by mutations which disrupted base pairing in the vicinity of the putative PTB binding site and also abolished internal initiation, with both properties restored by the compensating mutations that restored base pairing (JANG and WIMMER 1990). Apart from this one example, there is virtually no direct evidence that any of the other cross-linkable proteins actually play any role in internal initiation, and one cannot help wonder if the veritable deluge of recent publications which merely catalogue the cross-linkable proteins risk discrediting the whole field in the eyes of the outside world.

One approach to functional tests for *trans*-acting factors is to attempt to immunodeplete a cell-free extract of the putative factor. Immunodepleting HeLa cell extracts of PTB or the autoantigen La resulted in a selective loss of IRES-driven translation, but this activity could not be restored by addition of recombinant protein (HELLEN et al. 1993; Belsham et al., this volume). The conclusion drawn was that the functional component must be a complex of PTB or La with other proteins, rather than singular PTB (or La), although strictly speaking the data do not exclude the possibility that PTB and La are entirely irrelevant and it is only the putative associated proteins that are responsible for IRES-dependent translation initiation. However, using an RNA affinity column approach we have been able to deplete reticulocyte lysates of PTB, with the outcome that the efficiency of translation of the upstream (cap-dependent) cistron was unperturbed, but there was no translation of the downstream cistron driven by the EMCV IRES. Addition of recombinant PTB at physiologically relevant concentrations (10–12 μg/ml) completely restored the capacity for IRES-driven translation (A. KAMINSKI, S.L. HUNT and R.J. JACKSON, in preparation).

Another functional test for factors required for internal initiation is based on the fact that translation driven by the poliovirus or rhinovirus IRES in a reticulocyte lysate is inefficient unless components from HeLa, Krebs II ascites or L-cells are added (DORNER et al. 1984; PHILLIPS and EMMERT 1986; SVITKIN et al. 1988; BORMAN et al. 1993). One of these copurifies with PTB through four columns (S.L. Hunt and R.J. Jackson, unpublished observations). The other is largely if not entirely free of

PTB, but copurifies with a 97 kDa protein that is specifically cross-linkable to the rhinovirus IRES element (BORMAN et al. 1993). Neither activity copurifies with the auto-antigen La, or contains significant amounts of La (S.L. Hunt and R.J. Jackson, unpublished observations), which has been also claimed to stimulate poliovirus IRES utilisation in reticulocyte lysates (MEEROVITCH et al. 1993; SVITKIN et al. 1994). It is important to appreciate that this claim is not based on the approach of systematically purifying the HeLa cell activity which stimulates the ability of reticulocyte lysates to carry out translation dependent on the poliovirus IRES. Rather, a 52 kDa HeLa cell protein which binds in UV cross-linking assays to a particular stem-loop at the 3' end of the poliovirus IRES (MEEROVITCH et al. 1989) was identified as La, and as a consequence La was then tested for an influence on IRES-driven translation (MEEROVITCH et al. 1993). Recombinant La does indeed stimulate poliovirus RNA translation in reticulocyte lysates (MEEROVITCH et al. 1993; SVITKIN et al. 1994; Belsham et al., this volume), but only at very high concentrations, many fold higher than would be present in a mixed HeLa/rabbit reticulocyte cell-free system capable of efficient translation of poliovirus RNA. One possible explanation is that these high concentrations of La in these assays are actually inhibiting the initiation at incorrect sites which occurs when poliovirus RNA is translated in the unsupplemented reticulocyte lysate (DORNER et al. 1984; PHILLIPS and EMMERT 1986); as a consequence of this inhibition, initiation at the correct site might be stimulated by default.

Although PTB binds to all picornavirus IRESs in UV cross-linking assays, the entero/rhinovirus IRESs seem to bind a wider spectrum of proteins than the cardio/aphthovirus IRESs. In addition the pattern of proteins which are cross-linkable to other IRESs, such as that of BiP mRNA or hepatitis C virus (HCV) RNA, appears quite different and does not seem to include PTB as a strongly labelled protein. Moreover HCV IRES function does not seem to be impaired in the PTB-depleted lysate described above which is defective for EMCV IRES utilisation (J.E. Reynolds, A. Kaminski and R.J. Jackson, unpublished observations). If it is indeed the case that internal initiation on the HCV IRES, and perhaps other non-picornavirus IRESs, really does not require PTB, then the idea that PTB is an essential catalyst of the internal initiation process, in much the same way as the canonical initiation factors catalyse cap-dependent initiation, must be called into question. It is possible that the function of PTB is more to promote the correct folding of the RNA higher order structure of certain IRESs.

10 The Evolution and Possible Advantages of Internal Ribosome Entry Segments

Where did the animal picornavirus IRESs originate from? Obviously it is pertinent to look at the picorna-like viruses of plants and insects. Plant comoviruses, which show homology with animal picornaviruses in their capsid structures, gene order, and the covalently linked protein rather than a cap at the 5' end, have two RNA species: translation of the B RNA of the type member CPMV is initiated at the 5' proximal AUG codon at nt 207, and translation of the M RNA component at the second to fourth AUG codons (at nt 161, 512 and 524), whilst the out-of-frame 5' proximal AUG at nt 115 is presumed to be silent, perhaps because of its poor context. It has been claimed that the segment from nt 164 to nt 512 of the M RNA can direct internal initiation of translation in dicistronic mRNA assays in both the reticulocyte lysate and the wheat germ system (VERVER et al. 1991; THOMAS et al. 1991), although the efficiency was very low in some of these assays (THOMAS et al. 1991; SCHEPER et al. 1992). Moreover, the result could not be repeated in animal cell transfection assays (BELSHAM and LOMONOSOFF 1991). It seems possible that the in vitro result was due to the capricious or illegitimate internal initiation exhibited by the reticulocyte lysate, as discussed above, which may perhaps also occur in the wheat germ system. A clear difference between 5' NCRs of animal picornaviruses and comoviruses is that, whereas the former have AUG triplets at about the expected statistical (random) frequency, there appears to have been selection against such upstream AUGs in the comoviruses, even though CPMV M RNA does have one such AUG (at nt 115) which is not absolutely conserved. The long 5' NCR of the comovirus RNAs are generally U-rich (~ 35%–40%) and G-poor (12%–15%) and thus are probably not highly structured. In the absence of strong evidence to the contrary, it seems reasonable to conclude that the comovirus RNAs are not translated exclusively by internal initiation and that the predominant if not the sole mechanism is 5' end-dependent scanning, which because of the nature of the 5' NCR is not perceptibly influenced by a 5' cap structure.

It would be interesting to know whether the insect viruses such as cricket paralysis virus, which are thought to resemble the animal picornaviruses in many respects, have a 5' NCR closer to that of the animal picornaviruses or to the plant picorna-like viruses. This issue does not appear to have been addressed.

If the animal picornaviruses have acquired these highly structured IRESs relatively late in evolution, where did they come from? One possibility is that they arose by reiteration of a previously shorter 5' NCR and then mutation towards the present day IRES structure. However, it is difficult to see how the virus evolved successfully through the intermediate stages, when the 5' NCR would have been too structured to allow efficient 5' end-dependent scanning, yet had not developed to the point where internal ribosome entry is efficient. An alternative explanation is that the animal picornaviruses acquired their IRES sequences from cellular RNAs via some type of recombination event. In this respect it is interesting to note recent reports of poliovirus acquiring a short length of ribosomal RNA sequence (CHARINI et al. 1994) and pestivirus acquiring cellular ubiquitin mRNA coding sequences (MEYERS et al. 1991) by what would appear to be a copy choice recombination.

To date, there are a disproportionately large number of examples of internal initiation amongst viral RNAs rather than cellular, and although most viral examples are uncapped genomic RNAs, there is one case of a capped tricistronic mRNA translated by internal initiation (LIU and INGLIS 1992). Of course this preponderance of viral examples may be fortuitous in the sense that it is easier to spot likely candidates amongst viruses. In the absence of any systematic screening method to select or detect cellular examples in cDNA libraries, these can only be discovered on an individual trial and error basis. Nevertheless it is pertinent to ask why this mode of initiation is relatively common amongst viral RNAs. One suggestion relates to the fact that the extreme 5' end of viral RNAs has other functions, notably as the repository of signals for RNA replication, which might demand secondary structure (ANDINO et al. 1993) of sufficient stability to block 5' end-dependent scanning. However, this is clearly not an absolute rule since some positive strand RNA virus genomes, such as tobacco mosaic virus and the alphaviruses, have evolved so as to allow efficient RNA replication whilst maintaining a fairly unstructured 5' NCR that is not only permissive but is actually highly favourable to cap-dependent translation by a ribosome scanning mechanism (SLEAT et al. 1988).

In those viral RNAs with an IRES element, it is generally assumed that the RNA replication signals and the IRES element lie in distinct nonoverlapping domains of the 5' NCR. Thus with EMCV and FMDV RNAs, all the RNA replication signals are assumed to lie upstream of the poly(C) tract, and all the signals necessary for internal initiation are supposedly downstream of the tract. Likewise, in the rhino/enteroviruses the division between the upstream RNA replication signals (ANDINO et al. 1993) and the IRES is generally thought to be at around nt 100, where bovine enterovirus has a large insert and HRV-14 a small insert. The 5' boundary of the poliovirus IRES was in fact mapped as lying downstream of this point (NICHOLSON et al. 1991). However, this view of the RNA replication signals and the IRES as distinct nonoverlapping and noninteracting elements may be oversimplistic. It has been reported that mutations in the first 100 nt of the poliovirus 5' NCR, outside the region normally considered the IRES, can reduce the efficiency of expression of a linked luciferase reporter cistron in vivo (SIMOES and SARNOW 1991).

Conversely, there has been some recent evidence that the IRES element of polioviruses actually contains signals important for RNA replication. Until recently it has been impossible to detect this type of overlap, as IRES mutations which compromise viral RNA translation necessarily result in an RNA replication defect since some of the nonstructural viral proteins required for RNA replication can only function in cis (KIRKEGAARD 1992). This problem can be overcome by the use of viruses containing two IRES elements; the poliovirus IRES, in which mutations may be introduced, drives the synthesis of only the P1 capsid precursor, whilst a downstream EMCV IRES drives the expression of all the nonstructural proteins, which are therefore competent for replication of the template RNA strand in cis (MOLLA et al. 1992). This approach has shown that some poliovirus IRES mutants are not merely deficient in P1 translation, but also show severe RNA replication defects (BORMAN et al. 1994). A counterargument to the idea that some RNA replication signals might lie within the IRES element is provided by the fact that a poliovirus construct (with a single IRES), in which the poliovirus IRES has been replaced by that of EMCV, gives viable virus (ALEXANDER et al. 1994). This argues that if there are RNA replication signals within the IRES, they are somewhat surprisingly interchangeable between poliovirus and EMCV whose IRESs show only a rather limited homology (JACKSON et al. 1994).

An overlap between the translation initiation and the RNA replication signals is intriguing as it may offer a solution to one of the conundrums of positive strand RNA virus replication: how is the complementary (negative) RNA strand synthesised in the face of a wave of ribosomes moving in the opposite direction? Since the time when this problem was first recognised many years ago, it has become if anything more acute with the growing appreciation that an elongating ribosome is very efficient in displacing certainly any complementary RNA or DNA annealed to the coding region (MINSHULL and HUNT 1986; LINGENBACH and DOBBERSTEIN 1988) and probably also efficient at displacing proteins bound to the coding region. One solution to this problem is to suppose that before negative strand RNA synthesis starts initiation of translation is temporarily inhibited, a mechanism which would presumably require that the replication complex positioned at the 3' end of the positive (messenger) RNA strand should engage the IRES in some way. Hence the IRES may appear to include some signals essential for RNA replication, which may provide the explanation for the RNA replication defect of some poliovirus IRES mutants even in the background of a viral RNA where the nonstructural proteins are expressed from a separate and active IRES (BORMAN et al. 1994). Such a model carries the implication that it would be specifically the negative RNA strand synthesis that is inhibited by such poliovirus IRES mutations, a prediction which is so far untested.

11 Concluding Remarks

There are several important issues that have surfaced in this review and which need to be addressed in the near future. One is whether STNV RNA and the HSP mRNAs are really translated by an internal initiation mechanism or by an extreme cap-independent variant of the 5' end-dependent scanning mechanism. With respect to internal initiation, more data are needed on the mechanism and the identity and roles of the essential cis-acting RNA elements and the trans-acting protein factors, including the question of a possible role for eIF-4 in this mode of initiation. These questions need to be addressed not only regarding the two main types of picornavirus IRES, but also for other IRESs such as those in HCV RNA, BiP mRNA and the antennapedia mRNA of Drosophila (MACEJAK and SARNOW 1991; OH et al. 1992; Iizuka et al., this volume; Wang and Siddiqui, this volume), which at first glance show no resemblance to any picornavirus IRES either in their RNA sequence or structure, or in the IRES-binding proteins. Yet it seems instinctively quite improbable that distinct mechanisms of internal initiation are operative on the picornavirus IRESs and the non-picornavirus IRESs. It is extraordinary enough that there should be two apparently quite different modes of initiation, internal initiation and 5' end-dependent scanning. Surely, all examples of internal initiation must exhibit some common features both at the level of the cis-acting RNA elements and the trans-acting factors, and in turn one would expect to find some shared underlying features of internal initiation and cap-dependent initiation. One of the greatest challenges for the future is to find these common features and to work towards some form of a unified model, in which the two mechanisms would be regarded as variants of each other rather than as entirely distinct. Another major challenge is to develop screening or selection methods that may be applied to cDNA libraries to identify cellular mRNAs with a functional IRES, rather than the present day approach of trial and error tests on each individual mRNA.

Acknowledgments. We would like to thank past and present collaborators, and other members of our group, for all their input which has been instrumental in developing the ideas put forward here: Andy Borman, John Hershey, Michael Howell, Kathie Kean, Carola Lempke, Anne MacBride, Sue Milburn, Jim Patton, Bert Semler and Eckard Wimmer. Work carried out in the authors' laboratory was supported by a grant from the Wellcome Trust.

References

Agol VI (1991) The 5'-untranslated region of picornaviral genomes. Adv Virus Res 40: 103–180

Alexander L, Lu HH, Wimmer E (1994) Polioviruses containing picornavirus type 1 and/or type 2 internal ribosomal entry site elements: genetic hybrids and the expression of a foreign gene. Proc Natl Acad Sci USA 91: 1406–1410

Andino R, Rieckhof GE, Achacoso PL, Baltimore D (1993) Poliovirus RNA synthesis utilizes an RNP complex formed around the 5'-end RNA. EMBO J 12: 3587–3598

Anthony DD, Merrick WC (1991) Eukaryotic initiation factor (eIF)-4F. Implications for a role in internal initiation of translation. J Biol Chem 266: 10218–10226

Belsham GJ (1992) Dual initiation sites of protein synthesis on foot-and-mouth disease virus RNA are selected following internal entry and scanning of ribosomes in vivo. EMBO J 11: 1106–1110

Belsham GJ, Lomonosoff GP (1991) The mechanism of translation of cowpea mosaic virus middle component RNA: no evidence for internal initiation from experiments in an animal cell transient expression system. J Gen Virol 72: 3109–3113

Bergmann JE, Lodish HF (1979) Translation of capped and uncapped vesicular stomatitis virus and reovirus mRNAs. J Biol Chem 254: 459–468

Bonneau A-M, Sonenberg N (1987a) Involvement of the 24kd cap-binding protein in regulation of protein synthesis in mitosis. J Biol Chem 262: 11134–11139

Bonneau A-M, Sonenberg N (1987b) Proteolysis of p220 component of CBP complex is not sufficient for complete inhibition of host protein synthesis after poliovirus infection. J Virol 61: 986–991

Borman A, Jackson RJ (1993) Initiation of translation of human rhinovirus RNA: mapping the internal ribosomal entry site. Virology 188: 685–696

Borman A, Howell MT, Patton JG, Jackson RJ (1993) The involvement of a spliceosome component in internal initiation of human rhinovirus RNA translation. J Gen Virol 74: 1775–1788

Borman AM, Deliat FG, Kean KM (1994) Sequences within the poliovirus internal ribosome entry segment control viral RNA synthesis. EMBO J 13: 3149–3157

Buckley B, Ehrenfeld E (1987) The cap-binding proptein complex in uninfected and poliovirus infected HeLa cells. J Biol Chem 262: 13599–13606

Carrington JC, Freed DD (1990) Cap-independent enhancement of translation by a plant potyvirus 5' nontranslated region. J Virol 64: 1590–1597

Castrillo JL, Carrasco L (1987) Adenovirus late protein synthesis is resistant to the inhibition of translation induced by poliovirus. J Biol Chem 262: 7328–7334

Charini WA, Todd S, Gutman GA, Semler BL (1994) Transduction of a human RNA sequence by poliovirus. J Virol 68: 6547–6552
Chu L-Y, Rhoads RE (1978) Translational regulation of the 5'-terminal 7-methylguanosine of globin messenger RNA as a function of ionic strength. Biochemistry 17: 2450–2455
Danthine X, Seurinck J, Meulewaeter F, van Montagu M, Cornelissen M (1993) The 3' untranslated region of satellite tobacco necrosis virus RNA stimulates translation in vitro Mol Cell Biol 13: 3340–3349
De Benedetti A, Joshi-Barve S, Rinker-Schaeffer C, Rhoads RE (1991) Expression of antisense RNA against initiation factor eIF-4E mRNA in HeLa cells results in lengthened cell division times, diminished translation rates, and reduced levels of both eIF-4E and the p220 component of eIF-4F. Mol Cell Biol 11: 5435–5445
Devaney MA, Vakharia VN, Lloyd RE, Ehrenfeld E, Grubman MJ (1988) Leader protein of foot-and-mouth disease virus is required for cleavage of the p220 component of the cap-binding protein complex. J Virol 62: 4407–4409
Dolph PJ, Racaniello V, Villamarin A, Palladion F, Schneider RJ (1988) The adenovirus tripartite leader eliminates the requirement for cap binding protein during translation initiation. J Virol 62: 2059–2066
Dolph PJ, Huang J, Schneider RJ (1990) Translation by the adenovirus tripartite leader: elements which determine independence from cap-binding protein complex. J Virol 64: 2669–2677
Dorner AJ, Semler BL, Jackson RJ, Hanecak R, Duprey E, Wimmer E (1984) In vitro translation of poliovirus RNA: utilization of internal initiation sites in reticulocyte lysate. J Virol 50: 507–514
Drummond DR, Armstrong J, Colman A (1985) The effect of capping and polyadenylation on the stability, movement and translation of synthetic messenger RNAs in Xenopus oocytes. Nucleic Acids Res 13: 7375–7394
Edery I, Lee KAW, Sonenberg N (1984) Functional characterization of eukaryotic mRNA cap binding protein complex: effects on translation of capped and naturally uncapped RNAs. Biochemistry 23: 2456–2462
Etchison D, Hansen J, Ehrenfeld E, Edery I, Sonenberg N, Milburn SC, Hershey JWB (1984) Demonstration in vitro that eucaryotic initiation factor 3 is active but that a cap-binding protein complex is inactive in poliovirus-infected HeLa cells. J Virol 51: 832–837
Fletcher L, Corbin SD, Browning KG, Ravel JM (1990) The absence of a m^7G cap on β-globin mRNA and alfalfa mosaic virus RNA 4 increases the amounts of initiation factor 4F required for translation. J Biol Chem 265: 19582–19587
Futterer J, Kiss-Laszlo Z, Hohn T (1993) Nonlinear ribosome migration on cauliflower mosaic virus 35S RNA. Cell 73: 789–802
Gallie DR (1991) The cap and the poly(A) tail function synergistically to regulate mRNA translational efficiency. Genes Dev 5. 2108–2116
Gehrke L, Auron PE, Quigley GJ, Rich A, Sonenberg N (1983) 5'-conformation of a capped alfalfa mosaic virus ribonucleic acid 4 may reflect its independence of the cap structure or of cap binding protein for efficient translation. Biochemistry 22: 5157–5164
Gerstel B, Tuite MF, McCarthy JE (1992) The effects of 5' capping, 3'polyadenylation and leader composition upon the transational stability of mRNA in a cell-free extract derived from the yeast Saccharomyces cerevisiae. Mol Microbiol 6: 2339–2348
Gmyl AP, Pilipenko EV, Maslova SV, Belov GA, Agol VI (1993) Functional and genetic plasticities of the poliovirus genome: quasi-infectious RNAs modified in the 5'-untranslated region yield a variety of pseudo-revertants. J Virol 67: 6309–6316
Haller AA, Semler BL (1992) Linker scanning mutagenesis of the internal ribosome entry site of poliovirus RNA. J Virol 66: 5075–5086
Hambridge SJ, Sarnow P (1991) Terminal 7-methylguanosine cap structure on the normally uncapped 5'-noncoding region of poliovirus mRNA inhibits its translation in mammalian cells. J Virol 65: 6312–6315
Hellen CUT, Witherell GW, Schmid M, Shin SH, Pestova TV, Gil A, Wimmer E (1993) The cellular polypeptide p57 that is required for translation of picornavirus RNA by internal ribosome entry is identical to the nuclear pyrimidine-tract binding protein. Proc Natl Acad Sci USA 90: 7642–7646
Herson D, Schmidt A, Seal SN, Marcus A, van Vloten-Doting L (1979) Competitive mRNA translation in an in vitro system from wheat germ. J Biol Chem 254: 8245–8249
Huang J, Schneider RJ (1991) Adenovirus inhibition of cellular protein synthesis involves inactivation of cap binding protein. Cell 65: 271–280
Hunt SL, Kaminski A, Jackson RJ (1993) The influence of viral coding sequences on the efficiency of internal initiation of translation of cardiovirus RNAs. Virology 197: 801–807

Jackson RJ (1991) Potassium salts influence the fidelity of mRNA translation in rabbit reticulocyte lysates: unique features of encephalomyocarditis virus RNA translation. Biochim Biophys Acta 1088: 345–358

Jackson RJ, Howell MT, Kaminski A (1990) The novel mechanism of initiation of picornavirus RNA translation. Trend Biochem Sci 15: 477–483

Jackson RJ, Hunt SL, Gibbs CL, Kaminski A (1994) Internal initiation of translation of picornavirus RNAs. Mol Biol Rep 19: 147–159

Jang SK, Wimmer E (1990) Cap-independent translation of encephalomyocarditis virus RNA; structural elements of the internal ribosomal entry site and involvement of a cellular 57-kD RNA-binding protein. Genes Dev 4: 1560–1572

Jang SK, Kräusslich H-G, Nicklin MJH, Duke GM, Palmenberg AC, Wimmer E (1988) A segment of the 5' nontranslated region of encephalomyocarditis virus RNA directs internal entry of ribosomes during in vitro translation. J Virol 62: 2636–2643

Jia X-Y, Scheper G, Brown D, Updike W, Harmon S, Richards D, Summers D, Ehrenfeld E (1991) Translation of hepatitis A virus RNA in vitro: aberrant internal initiations influenced by 5' noncoding region. Virology 182: 712–722

Joshi B, Yan R, Rhoads RE (1994) In vitro synthesis of human protein synthesis initiation factor eIF-4γ and its localization on 43S and 48S initiation complexes. J Biol Chem 269: 2048–2055

Joshi-Barve S, Rychlik W, Rhoads RE (1990) Alteration of the major phosphorylation site of eukaryotic protein synthesis initiation factor 4E prevents its association with the 48S initiation complex. J Biol Chem 265: 2979–2983

Joshi-Barve S, De Benedetti A, Rhoads RE (1992) Preferential translation of heat shock mRNAs in HeLa cells deficient in protein synthesis initiation factors eIF-4E and eIF-4γ J Biol Chem 267: 21038–21043

Kaminski A, Howell MT, Jackson RJ (1990) Initiation of encephalomyocarditis virus RNA translation: the authentic initiation site is not selected by a scanning mechanism. EMBO J 9: 3753–3759

Kaminski A, Belsham GJ, Jackson RJ (1994a) Translation of encephalomyocarditis virus RNA; parameters influencing the selection of the internal initiation site. EMBO J 13: 1673–1681

Kaminski A, Hunt SL, Gibbs CL, Jackson RJ (1994b) Internal initiation of mRNA translation in eukaryotes. In: Setlow J (ed) Genetic engineering. principles and methods, vol 16. Plenum, New York, pp 115–155

Kemper B, Stolarsky L (1977) Dependence on potassium concentration of the inhibition of the translation of messenger ribonucleic acid by 7-methylguanosine 5'-monophosphate. Biochemistry 16: 5676–5680

Kirkegaard K (1992) Genetic analysis of picornaviruses. Curr Opin Genet Dev 2: 64–70

Koromilas AE, Lazaras-Karatzas A, Sonenberg N (1992) mRNAs containing extensive secondary structure in their 5' non-coding region translate efficiently in cells overexpressing initiation factor eIF-4E EMBO J 11: 4153–4158

Kozak M (1978) How do eukaryotic ribosomes select initiation regions in messenger RNA? Cell 15: 1109–1123

Kozak M (1989) The scanning model for translation: an update. J Cell Biol 108: 229–241

Kozak M (1991) A short leader sequence impairs the fidelity of initiation by eukaryotic ribosomes in vitro. Gene Expr 1: 111–116

Kuge S, Nomoto A (1987) Construction of viable deletion and insertion mutants of the Sabin strain of type 1 poliovirus: function of the 5' noncoding sequence in viral replication. J Virol 61: 1478–1487

Lamphear BJ, Panniers R (1991) Heat shock impairs the interaction of cap binding protein complex with 5' mRNA cap. J Biol Chem 266: 2789–2794

Liebig H-D, Ziegler E, Yan R, Hartmuth, Klump H, Kowlaski H, Blass D, Sommergruber W, Frasel L, Lamphear B, Rhoads RE, Kuechler E, Skern T (1993) Purification of two picornaviral 2A proteinases: interaction with eIF-4γ and influence on in vitro translation. Biochemistry 32: 7581–7588

Lindquist S, Petersen R (1991) Selective translation and degradation of heat shock messenger RNAs in Drosophila. Enzyme 44: 147–166

Lingenbach K, Dobberstein B (1988) An extended RNA/RNA duplex structure within the coding region of mRNA does not block translation elongation. Nucleic Acids Res 16. 3405–3414

Liu DX, Inglis SC (1992) Internal entry of ribosomes on a tricistronic mRNA encoded by infectious bronchitis virus. J Virol 66: 6143–6154

Lodish HF, Rose JK (1977) Relative importance of 7-methylguanosine in ribosome binding and translation of vesicular stomatitis virus mRNA in wheat germ and reticulocyte cell-free systems. J Biol Chem 252: 1181–1188

Macejak DG, Sarnow P (1991) Internal initiation of translation mediated by the 5' leader of a cellular mRNA. Nature 353: 90–94
McGarry TJ, Lindquist S (1985) The preferential translation of Drosophila hsp70 mRNA requires sequences in the untranslated leader. Cell 42: 903–911
McMillan JP, Singer MX (1993) Translation of the human LINE-1 element, L1Hs. Proc Natl Acad Sci USA 90: 11533–11537
Meerovitch K, Pelletier J, Sonenberg N (1989) A cellular protein that binds to the 5'-noncoding region of poliovirus RNA: implications for internal translation initiation. Genes Dev 3: 1026–1034
Meerovitch K, Nicholson R, Sonenberg N (1991) In vitro mutational analysis of cis-acting RNA translational elements within the poliovirus type 2 5' untranslated region. J Virol 65: 5895-5901
Meerovitch K, Svitkin YV, Lee HS, Lejbkowicz F, Kenan DJ, Chan EKL, Agol VI, Keene JD, Sonenberg N (1993) La autoantigen enhances and corrects aberrant translation of poliovirus RNA in reticulocyte lysate. J Virol 67: 3798–3807
Meyers G, Tautz N, Dubovi EJ, Theil H-J (1991) Viral cytopathogenicity correlated with integration of ubiquitin-coding sequences. Virology 180: 602–616
Minshull J, Hunt T (1986) The use of single-stranded DNA and RNase H to promote quantitative 'hybrid arrest of translation' of mRNA/DNA hybrids in reticulocyte lysate cell-free translations. Nucleic Acids Res 14: 6433–6451
Molla A, Jang SK, Paul AV, Reuer Q, Wimmer E (1992) Cardioviral internal ribosomal entry site is functional in a genetically engineered dicistronic poliovirus. Nature 356: 255–257
Moore MJ, Sharp PA (1992) Site-specific modification of pre-mRNA; the 2'-hydroxyl groups at the splice sites. Science 256: 992–997
Muñoz A, Alonson MA, Carrasco L (1984) Synthesis of heat-shock proteins in HeLa cells: inhibition by virus infection. Virology 137: 150–159
Nicholson R, Pelletier J, Le S-Y, Sonenberg N (1991) Structural and functional analysis of the ribosome landing pad of poliovirus type 2: in vivo translation studies. J Virol 65: 5886–5894
Oh SK, Scott MP, Sarnow P (1992) Homeotic gene antennapedia messenger RNA contains 5'-non-coding sequences that confer translational initiation by internal ribosome binding. Genes Dev 6: 1643–1653
Pause A, Methot N, Svitkin Y, Merrick WC, Sonenberg (1994) Dominant negative mutants of mammalian translation initiation factor eIF-4A define a critical role for eIF-4F in cap-dependent and cap-independent initiation of translation. EMBO J 13: 1205–1215
Peabody DS (1989) Translation initiation at non-AUG triplets in mammalian cells. J Biol Chem 264: 5031–5035
Pelletier J, Kaplan G, Racaniello VR, Sonenberg N (1988a) Cap-independent translation of poliovirus mRNA is conferred by sequence elements within the 5' noncoding region. Mol Cell Biol 8: 1103–1112
Pelletier J, Flynn ME, Kaplan G, Racaniello V, Sonenberg N (1988b) Mutational analysis of upstream poliovirus AUG codons of poliovirus RNA. J Virol 62: 4486–4492
Perez L, Carrasco L (1991) Lack of direct correlation between p220 cleavage and the shut off of host translation after poliovirus infection. Virology 189: 178–186
Pestova TV, Hellen CUT, Wimmer E (1991) Translation of poliovirus RNA; role of an essential cis-acting oligopyrimidine element within the 5' nontranslated region and involvement of a cellular 57-kilodalton protein. J Virol 65: 6194–6204
Phillips BA, Emmert A (1986) Modulation of the expression of poliovirus proteins in reticulocyte lysates. Virology 148: 255–267
Pilipenko EV, Gmyl AP, Maslova SV, Svitkin YV, Sinyakov AN, Agol VI (1992) Prokaryotic-like cis elements in the cap-independent internal initiation of translation on picornavirus RNA. Cell 68: 119–131
Rhoads RE, Joshi-Barve S, Rinker-Schaeffer C (1993) Mechanism of action and regulation of protein synthesis initiation factor 4E: effects on mRNA discrimination, cellular growth rate, and oncogenesis. Prog Nucleic Acid Res Mol Biol 46: 183–219
Rozen F, Edery I, Meerovitch K, Dever TE, Merrick WC, Sonenberg N (1990) Bidirectional RNA helicase activity of eucaryotic initiation factors 4A and 4F. Mol Cell Biol 10: 1134–1144
Sarkar G, Edery I, Gallo R, Sonenberg N (1984) Preferential stimulation of rabbit α-globin mRNA translation by a cap-binding protein complex. Biochim Biophys Acta 783: 122–129
Sarnow P (1989) Translation of glucose regulated protein 78/immunoglobulin heavy chain binding protein mRNA is increased in poliovirus–infected cells at a time when cap-dependent translation of cellular mRNAs is inhibited. Proc Natl Acad Sci USA 86: 5795–5799

Scheper GC, Voorma HO, Thomas AAM (1992) Eukaryotic initiation factors-4E and -4F stimulate 5' cap-dependent as well as internal initiation of protein synthesis. J Biol Chem 267: 7269–7274
Simoes EA, Sarnow P (1991) An RNA hairpin at the extreme 5' end of the poliovirus RNA genome modulates viral translation in human cells. J Virol 65: 913–921
Sleat DE, Hull R, Turner PC, Wilson TMA (1988) Studies on the mechanism of translational enhancement by the 5'-leader sequence of tobacco mosaic virus RNA. Eur J Biochem 175: 75–86
Sonenberg N (1988) Cap-binding proteins of eukaryotic messenger RNA: functions in initiation and control of translation. Prog Nucleic Acid Res Mol Biol 35: 173–207
Sonenberg N, Guertin D, Cleveland D, Trachsel H (1981) Probing the function of the eukaryotic 5' cap structure by using a monoclonal antibody directed against cap-binding proteins. Cell 27: 563–572
Sonenberg N, Guertin D, Lee KAW (1982) Capped mRNAs with reduced secondary structure can function in extracts from poliovirus infected cells. Mol Cell Biol 2: 1633–1638
Staehelin T, Trachsel H, Erni B, Boschetti A, Schreier MH (1975) The mechanism of initiation of mammalian protein synthesis. FEBS Symp 39: 309–323
Sun X-H, Baltimore D (1989) Human immunodeficiency virus tat-activated expression of poliovirus protein 2A inhibits mRNA translation. Proc Natl Acad Sci USA 86: 2143–2146
Svitkin YV, Pestova TV, Maslova SV, Agol VI (1988) Point mutations modify the response of poliovirus RNA to a translation initiation factor: a comparison of neurovirulent and attenuated strains. Virology 166: 394–404
Svitkin YV, Meerovitch K, Lee HS, Dholakia JN, Kenan DJ, Agol VI, Sonenberg N (1994) Internal translation initiation on poliovirus RNA: further characterization of La function in poliovirus translated in vitro. J Virol 68: 1544–1550
Thomas AAM, ter Haar E, Wellink J, Voorma HO (1991) Cowpea mosaic virus middle component RNA contains a sequence that allows internal binding of ribosomes and that requires eukaryotic initiation factor 4F for optimal translation. J Virol 65: 2953–2959
Thomas AAM, Scheper GC, Kleijn M, DeBoer M, Voorma HO (1992) Dependence of the adenovirus tripartite leader on the p220 subunit of eukaryotic initiation factor 4F during in vitro translation. Eur J Biochem 207: 471–477
Timmer RT, Benkowski LA, Schodin D, Lax SR, Metz AM, Ravel JM, Browning KS (1993a) The 5' and 3' untranslated regions of satellite tobacco necrosis virus RNA affect translational efficiency and dependence on 5' cap structure. J Biol Chem 268: 9404–9510
Timmer RT, Lax SR, Hughes DL, Merrick WC, Ravel JM, Browning KS (1993b) Characterisation of wheat germ protein synthesis initiation factor eIF-4C and comparison of eIF-4C from wheat germ and rabbit reticulocytes. J Biol Chem 268: 24863–24867
Tsukiyama-Kohara K, Iizuka N, Kohara M, Nomoto A (1992) Internal ribosome entry site within hepatitis C virus RNA. J Virol 66: 1476–1483
Verver J, Le Gall O, van Kammen A, Wellink J (1991) The sequence between nucleotides 161 and 512 of cowpea mosaic virus M RNA is able to support internal initiation of translation in vitro. J Gen Virol 72: 2339–2345
Wang C, Sarnow P, Siddiqui A (1993) Translation of human hepatitis C virus RNA in cultured cells is mediated by an internal ribosome-binding mechanism. J Virol 67: 3338–3344
Weber LA, Hickey ED, Baglioni C (1978) Influence of potassium salt concentration and temperature on inhibition of mRNA translation by 7-methylguanosine 5'-monophosphate. J Biol Chem 253: 178–183
Wodnar-Filipowicz A, Szczesna E, Zan-Kowalczewska M, Muthukrishnan S, Szybiak U, Legocki AB, Filipowicz W (1978) 5'-terminal 7-methylguanosine and mRNA function. Eur J Biochem 92: 69–80
Wyckoff EE, Lloyd RE, Ehrenfeld E (1992) Relationship of eukaryotic initiation factor 3 to poliovirus-induced p220 cleavage activity. J Virol 66: 2943–2951
Yan R, Rychlik W, Etchison D, Rhoads RE (1992) Amino acid sequence of the human protein initiation factor eIF-4γ. J Biol Chem 267: 23226–23231
Yoo BJ, Spaete RR, Gaballe AP, Selby M, Houghton M, Han JH (1992) 5' end-dependent translation initiation of hepatitis C viral RNA and the presence of putative positive and negative translation control elements within the 5' untranslated region. Virology 191: 889–899
Zapata JM, Maroto FG, Sierra JM (1991) Inactivation of mRNA cap-binding protein complex in Drosophila melanogaster embryos under heat shock. J Biol Chem 266: 16007–16014

Translation of Encephalomyocarditis Virus RNA by Internal Ribosomal Entry

C.U.T. HELLEN[1] and E. WIMMER[2]

1 Introduction

The positive-sense genomic RNAs of picornaviruses such as encephalomyocarditis virus (EMCV) have been widely used in studies of translation, resulting in significant advances, such as identification of mammalian Met-tRNA (SMITH and MARCKER 1970) and recognition of the initiation factor eIF-4A (WIGLE and SMITH 1973; BLAIR et al. 1977). Analysis of picornavirus translation also revealed a fundamental difference between eukaryotic and prokaryotic mRNAs: initiation of translation is limited to a single 5' proximal site (JACOBSON and BALTIMORE 1968; SMITH 1973), indicating that picornavirus genomes, and by implication all other

[1]Department of Microbiology and Immunology, SUNY Health Sciences Center at Brooklyn, 450 Clarkson Avenue, Box 44, Brooklyn, NY 11203-2098, USA
[2]Department of Microbiology, State University of New York at Stony Brook, Stony Brook, NY11794-8621, USA

eukaryotic mRNAs, are monocistronic. The EMCV genome does indeed contain a single large open reading frame, but further studies of picornavirus mRNAs revealed fundamental differences from standard eukaryotic mRNAs (such as the absence of a 5' terminal capping group, and the presence of multiple AUG triplets and stable secondary structure upstream of the initiation codon) that proved incompatible with conventional models for the initiation of eukaryotic translation (KOZAK 1991). The discovery that EMCV initiation results from entry of ribosomes into an internal segment of the 5'NCR (JANG et al. 1988) has revitalized studies of EMCV translation, and its internal ribosome entry site (IRES) is now used both as a model for analysis of this novel mechanism of eukaryotic gene expression, and as a genetic element, for example in expression vectors, to promote cap-independent internal initiation of translation.This review shall focus on EMCV translation, but will also consider translation of other cardioviruses and of the related aphthoviruses and hepatoviruses.

The picornavirus family consists of a large number of animal viruses whose positive-sense RNA genomes have a similar genetic organization and share common structural properties. Their genomes are covalently linked to a small protein (VPg), 3'polyadenylated, have exceptionally long 5'NCRs (600–1500 nt) and encode a single polyprotein of approximately 250 kDa (STANWAY 1990). Picornaviruses are divided into five genera on the basis of biological and physical properties, and sequence analysis indicates a probable common ancestry (PALMENBERG 1989). These genera have been divided into two groups on the basis of primary sequences, and similarities in the secondary structures of their IRES elements (JACKSON et al. 1990; JANG et al. 1990). The genera Enterovirus (e.g., poliovirus) and Rhinovirus constitute the first group, and they thus have type 1 IRES elements. Type 2 IRES elements belong to members of the genera Aphthovirus (e.g., foot-and-mouth disease virus; FMDV), and Cardiovirus (e.g., EMCV). The secondary structure of Hepatovirus (e.g., hepatitis A virus; HAV) IRES elements resembles that of type 2 IRES elements, but in functional assays in vitro the HAV IRES is remarkably inefficient. Finally, echovirus 22(EV22), which may represent a sixth genus (HYPPIA et al. 1992) also appears to have type 2 IRES.

2 Characteristics of Encephalomyocarditis Virus Translation In Vivo and In Vitro

Initiation of EMCV translation is exceptionally efficient (JEN et al. 1978), and synthesis of viral proteins is apparent soon after infection (BUTTERWORTH et al. 1971). EMCV has a broad host range, and its RNA is translated efficiently in a variety of mammalian cell lines and even in Xenopus oocytes (LASKEY et al. 1972). Infection is associated with disaggregation of polysomes synthesizing cellular proteins and a gradual inhibition of host cell protein synthesis (DALGARNO et al. 1967; JEN and THACH 1982; LAWRENCE and THACH 1974). Translation of cellular mRNAs is probably inhibited because they have a lower affinity than EMCV RNA for initiation factors (BAGLIONI et al. 1978; GOLINI et al. 1976; HACKETT et al. 1978). Inhibition of host cell protein synthesis can be relieved by decreasing the concentration of monovalent ions in the extracellular medium; conversely, hypertonic conditions promote viral translation (ALONSO and CARRASCO 1981). Inhibition thus differs from the "shut-off" that results from poliovirus infection (JEN et al. 1980; MOSENKIS et al. 1985), which is associated with proteolysis of eIF-4Fγ (ETCHISON et al. 1982). However, EMCV RNA can be translated in poliovirus-infected cells, indicating that its initiation does not required eIF-4Fγ and is thus cap-independent (DETJEN et al. 1981). This conclusion is supported by recent studies involving dicistronic polioviruses (ALEXANDER et al. 1994; MOLLA et al. 1992).

EMCV RNA is translated efficiently in different eukaryotic cell-free lysates, including rabbit reticulocyte lysate (RRL) and lysates derived from chick embryo fibroblasts, HeLa cells, murine L cells, Krebs II and Ehrlich ascites carconima cells (GOLINI et al. 1976; MATHEWS and KORNER 1970; PELHAM 1978; PELHAM and JACKSON 1976; SVITKIN and AGOL 1978). This ability distinguishes EMCV translation from that of some other Picornaviruses, such as poliovirus (DORNER et al. 1984). However, EMCV RNA is not translated in wheat germ lysate (JEN and THACH 1982) or in the yeast *Saccharomyces cerevisiae* (EVSTAFIEVA et al. 1993) Optimal EMCV translation requires unusually high salt concentrations: protein synthesis occurs over a narrow range of Mg^{2+} concentrations, and is maximal at 4–5 m*M* (MATHEWS and KORNER 1970; SVITKIN and AGOL 1978). The optimum K^+ concentration (120 m*M*) is also high; stimulation of authentic product synthesis in RRLs was activated by KCl, or low concentrations of KSCN, but not by potassium acetate, although this salt is known to stimulate translation of cap-dependent mRNAs (JACKSON 1991a). EMCV translation is relatively insensitive to cap analogues such as m^7pG, indicating that formation of initiation complexes on Cardiovirus mRNA is cap-independent (CANAANI et al. 1976). Capping EMCV mRNAs had no effect on their translation efficiency, consistent with this observation (OUDSHOORN et al. 1990). Formation of initiation complexes on EMCV RNA is uniquely independent of ATP (JACKSON 1991b).

3 Primary Structure of the Encephalomyocarditis Virus 5'NCR

The 5'terminal of the EMCV RNA genome differs from that of cellular mRNAs in that it is neither capped nor does it carry unblocked 5'phosphates (FRISBY et al. 1976), but is instead covalently linked to a small virus-encoded protein (GOLINI et al. 1978; HRUBY and ROBERTS 1978; VARTAPETIAN et al. 1980) Removal of the VPg moiety from mengovirus RNA does not affect its ability to initiate translation (PEREZ-BERCOFF and GANDER 1978); similar results have been obtained using FMDV RNA (SANGAR et al. 1980). These properties are common to all picornaviruses (WIMMER 1982).

Determination of complete or partial nucleotide sequences of EV22 and over 35 aphthovirus and cardiovirus strains and isolates has revealed that their 5'NCRs are closely related and have several unusual properties, including their extreme length (710–1500 nt) and the presence of multiple AUG triplets (DUKE et al. 1992; ESCARMIS et al. 1992; HYPPIA et al. 1992; LAW and BROWN 1990; PRITCHARD et al. 1992; SANGER et al. 1987; SOSNOVTSEV et al. 1993; VARTAPETIAN et al. 1983; STANWAY 1990; ZIMMERMANN et al. 1994). A poly(C) tract of 60–420 nt is present near the 5' end of the genome in all FMDV isolates, in some Cardioviruses including EMCV, mengovirus and Maus-Elberfeld virus, but not Theiler's murine enphalomyocarditis virus (TMEV), and not in either EV22 or HAV (BLACK et al. 1979; BROWN et al. 1974; CHUMAKOV and AGOL 1976; CHUMAKOV et al. 1979; COLLER et al. 1990; ESCARAMIS et al. 1992; PORTER et al. 1974). The FMDV poly(C) tract is largely single-stranded and probably adopts an extended helical conformation (BRAHMS et al. 1967). A potentially unstructured pyrimidine-rich tract occurs at a similar position in all other picornavirus genera and is the site of extensive sequence microheterogeneity between isolates (BROWN et al. 1991; LE and ZUKER 1990; PÖYRY et al. 1992). Nucleotides upstream of the poly(C) tract are termed the "short" or "S" fragment and range in length from 140 nt in EMCV to 370 nt in FMDV. They are not required for translation of genomic FMDV and EMCV RNAs but are essential for infectivity and are probably cis-acting signals for genome replication. This observation suggests that the 5'NCR of these viruses may have a modular organization reflecting these two functions (SANGAR et al. 1980; CHUMAKOV et al. 1979).

EMCV and mengovirus 5'NCRs contain 758–859 nt, and those of FMDV strains 1200–1500 nt depending on the length of the poly(C) tract. There is less variation in the length of 5'NCR in viruses that do not contain such a tract; the TMEV, HAV and EV22 5'NCRs contains 1064 nt, 734 nt, and 709 nt respectively. Sequence identity between the 5'NCRs of EV22 and members of the cardiovirus and aphthovirus genera is largely restricted to the 450–550 nt preceding the initiation codon. This region in the FMDV 5'NCR is 40% identical to corresponding segments of both EMCV and TMEV (KÜHN et al. 1990), and sequence identity is even higher between members of the same genus. The sequences of TMEV and EMCV in this region are 68% identical (PEVEAR et al. 1987), and analysis of complete or partial FMDV 5'NCR sequences indicates that the 3'proximal 500 nt are 80% identical in 17 different strains. These long conserved segments of the 5'NCR upstream of the initiation codon correspond to the IRES (see Sect. 3). The 5'NCRs of Hepatoviruses are related to the 5'NCRs of these three picornavirus genera, although not closely (BROWN et al. 1991; JANG et al. 1990).

Sequence alignment of type 2 IRES elements reveals that regions of nucleotide sequence identity are localized into discrete blocks. Many of these motifs correspond to structural domains within the IRES, particularly to apical portions of hairpins (Figs.1–3), and interestingly, some appear also to be present in type 1 IRES elements. The presence of near-identical sequence motifs in all type 2 IRES elements suggests that they could be cis-acting elements involved in sequence-specific interactions with trans-acting factors. Sequence variation is also tightly clustered; most of the nonconserved residues in the FMDV IRES occur in just three locations: the central stem of domain I, domain K and the region between domain L and the initiation codon (Fig. 2). The role of these and similar RNA segments (such as domain G, and residues downstream of domain L in cardioviruses; Fig. 1) is unlikely to be sequence-dependent, but may instead be related to their length or structure.

Initiation of EMCV translation occurs at a single site (AUG_{834}) which was identified by matching sequences derived from the viral genome (PALMENBERG et al. 1984) and viral proteins translated from it in vitro and in vivo (Smith 1973; UGAROVA et al. 1984). However, all aphthoviruses and hepatoviruses, and many cardioviruses have dual initiation sites; those of HAV and FMDV are in the same reading frame and are separated by 3 nt, and 78 or 84 nt, respectively, whereas those of TMEV are 10 nt apart and are thus in overlapping reading frames (BECK et al. 1983; CLARKE et al. 1985; KONG and ROOS 1991; PRITCHARD et al. 1992; SANGAR et al. 1987; TESAR et al. 1991). Utilization of these codons is not strictly dependent on conventional context rules; for example, the first initiation codon in FMDV A_{10} is bypassed frequently even though it has a good context (AUU*AUG*A), whereas in other serotypes it has a less favourable context (e.g., UAA*AUG*G in type A24), but is nevertheless recognized efficiently (SANGAR et al. 1987; see a discussion by JANG et al. 1990).

4 Secondary and Tertiary Structure of the Encephalomyocarditis Virus 5'NCR

Initial investigation of the stricture of *Aphthovirus* and *Cardiovirus* 5'NCRs concentrated on the S fragment and the poly(C) tract (NEWTON et al. 1985; VARTAPETIAN et al. 1983). Nucleotides immediately downstream of the poly(C) tract in different FMDV isolates can potentially form three or four pseudoknots (CLARKE et al. 1987; ESCARMIS et al. 1992), and nucleotides immediately upstream of the EMCV poly(C) tract and its HAV equivalent can form two pseudoknots (BROWN et al. 1991; DUKE et al. 1992; WIMMER and MURDIN 1991). The function of these structures is unknown. Proposed tertiary interactions at the 3' end of Aphthovirus, Cardiovirus and Hepatovirus 5'NCRs involve conserved elements upstream and downstream of domain L (LE et al. 1993).

Efforts to determine the secondary structures of the remainder of these RNAs were revitalized by identification of IRES elements within FMDV, EMCV and TMEV 5'NCRs, and by demonstration of the functional importance of their integrity. The availability of a large set of nucleotide sequences has facilitated determination of potential secondary structures using computational and phylogenetic approaches (DUKE et al. 1992; PILIPENKO et al. 1989). Conclusions reached in this way have been supported by mutational studies (e.g.,JANG and WIMMER 1990) and by the analysis of RNA sensitivity to chemical and enzymatic modification (EVSTAFIEVA et al. 1991; PILIPENKO et al. 1989).

A consensus structure of the EMCV 5'NCR downstream of the poly(C) tract is shown in Fig. 1. Domains are named according to DUKE et al. (1992); the structure is based on analysis by these authors and by PILIPENKO et al. (1989). The majority of nucleotides are base-paired, forming a series of hairpins. A number of nucleotides are invariant in the sequences of 17 cardioviruses, and are concentrated in domains D, H, I, J, K and L. Most sequence variation takes the form of transitions and, to a lesser extent, compensatory double mutations, so that these changes cause little or no structural alteration. There are differences between the models developed by these two groups, in particular concerning the structure adopted by nt 393–403, the base-pairing between nt 459–518 and nt 600–670, the size and nucleotide constituents of domain L, and the possibility of structural motifs downstream of this domain. Phylogenetic comparison supports the possibility that nt 393–403 could form a small hairpin (Fig. 1), but provides little support for the existence of hairpins downstream of domain L. There are currently no compelling thermodynamic or phylogenetic arguments to discriminate between the two proposed structures of the central stem in domain I. Models have been developed for the structure of the FMDV IRES (PILIPENKO et al. 1989; Fig. 2) and the EV22 5'NCR (Fig. 3) that are closely related to the structure of the EMCV IRES; some similarities between the structures of these elements and the HAV IRES are also apparent (BROWN et al. 1991; LE et al. 1993).

It is notable that many of the nucleotides that are invariant in different EMCV, TMEV and mengovirus strains also occur in identical locations in the FMDV and EV22 5'NCRs. A likely implication of these observations is that type 2 IRES elements consist of a number of cis-acting elements assembled on an RNA "scaffold" of conserved architecture.

5 The Internal Ribosomal Entry Site

Translation of mRNAs containing segments of the EMCV 5'NTR linked to reporter genes confirmed suggestions (CHUMAKOV et al. 1979) that the 5'NCR segment downstream of the poly(C) tract was sufficient to direct efficient translation in vitro (KRÄUSSLICH et al. 1987; PARKS et al. 1986). The 500 nt between the poly(C) tract and the EMCV initiation codon contain ten AUG triplets (see Fig. 1), yet sequence analysis of polypeptides synthesized in vivo and in vitro indicated that only the 11th AUG triplet at nt 834 is a functional initiation codon (SMITH 1973; UGAROVA et al. 1984). These results cannot be readily reconciled with the "scanning" model for the initiation for translation (KOZAK 1991). A likely alternative explanation is that ribosomes are able to bypass these potential initiation codons and initiate translation at AUG_{834} by a scanning-independent mechanism. This hypothesis was confirmed by JANG and colleagues, who constructed plasmids for the transcription of bicistronic mRNAs, as depicted in Fig. 4, and found that a segment (nt 260–833) of the EMCV 5'NCR placed in the intercistronic region was

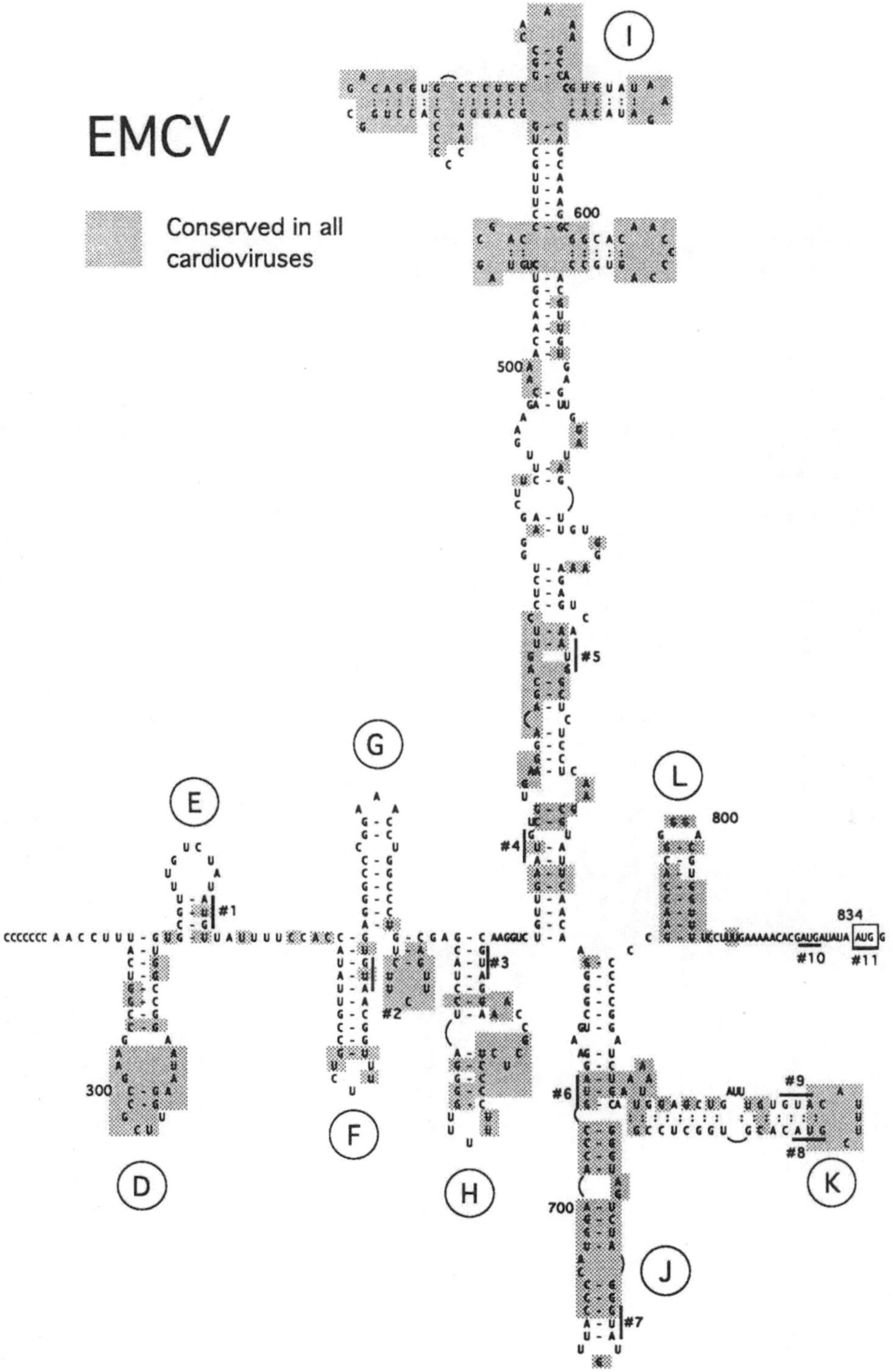

Fig. 1. (Legend see p. 40)

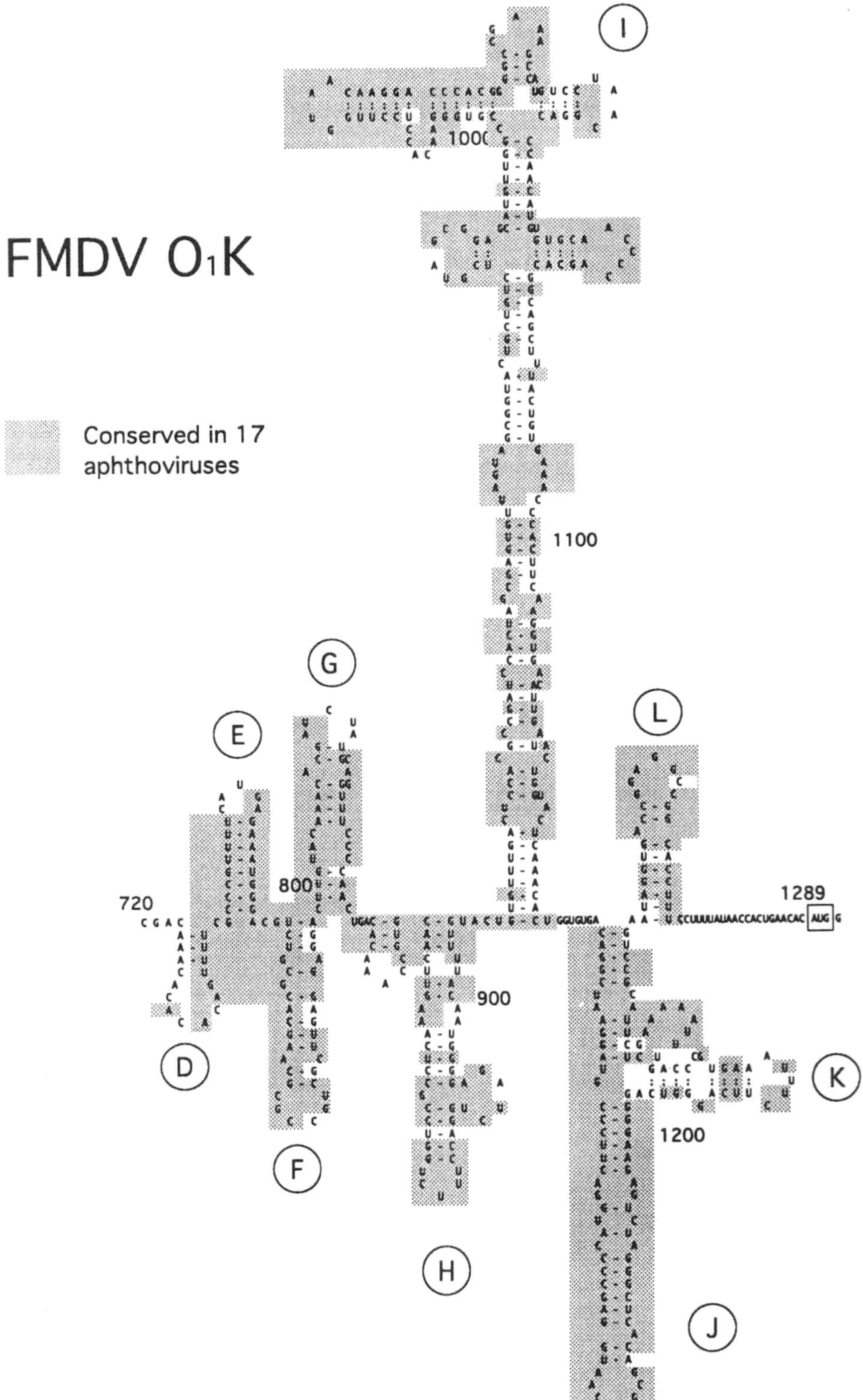

Fig. 2. (Legend see p. 40)

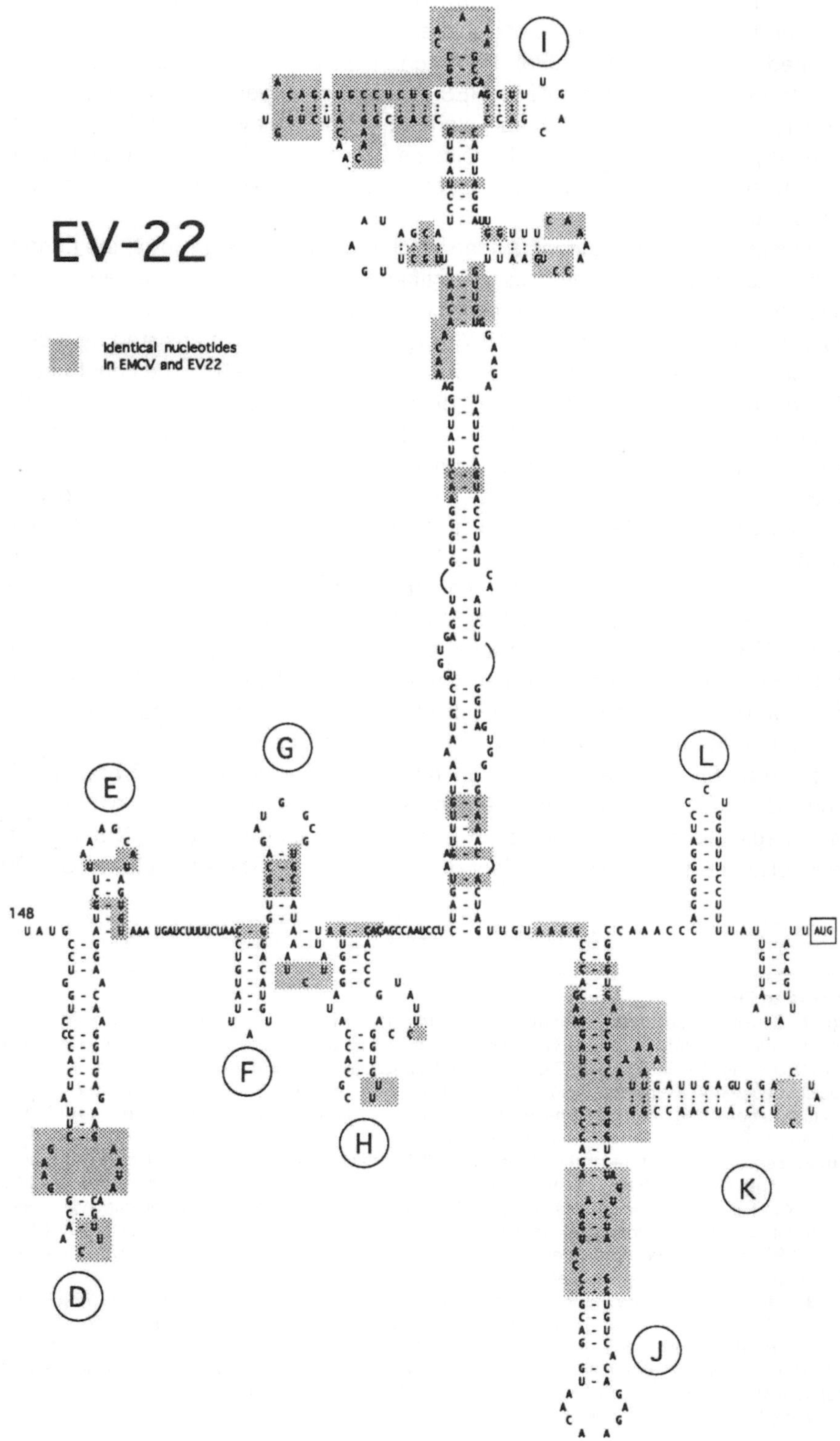

Fig. 3. (Legend see p. 40)

able to direct efficient translation of the downstream cistron in vitro and in vivo, even under circumstances in which translation of the upstream cistron was inhibited (JANG et al. 1988, 1989). This segment of the EMCV 5'NCR was termed the internal ribosome entry site or IRES. An identical experimental approach was used to identify a similar genetic element in the poliovirus 5'NCR, and subsequently, in the 5'NCRs of all other picornavirus genera (Belsham and Brangwyn 1990; BORMAN and JACKSON 1992; GLASS et al. 1993; KÜHN et al. 1990; PELLETIER and SONENBERG 1988).

Despite the overwhelming evidence for internal ribosomal entry, the phenomenon has been discounted as an artifact, caused either by degradation of dicistronic mRNAs in vitro or by their splicing/aberrant transcription in vivo, as a result of which monocistronic mRNA fragments would be translated by a conventional scanning mechanism (see, for example, KOZAK 1992). The construction of viable dicistronic polioviruses (Fig. 8), in which the open reading frame (ORF) was interrupted by insertion of the EMCV IRES (MOLLA et al. 1992; see Sect. 10) rendered such criticism redundant. Titration of this virus indicated that each virus particle had the propensity to initiate an infectious cycle (MOLLA et al. 1992). The viral genome must therefore have functioned with the internal IRES element intact.

Four lines of evidence suggest that the structural integrity of the EMCV IRES is essential for its function. First, hybridization of short oliogodeoxynucleotides to nt 309–338 of the 5'NCR had no effectr, and to nt 420–449 little effect on the ability of the 5'NCR to direct translation of monocistronic mRNAs in RRL. Hybridization of any of nine additional oligodeoxynucleotides to sequences between nt 450–854 inhibited translation severely (SHIH et al. 1987). Similar results were subsequently reported by SANKAR et al. (1989). Second, linker insertion at nt 577 and nt 777 reduced initiation 6- and 12-fold, respectively, whereas linker insertion at nt 506 had a negligible effect on translation (DUKE et al. 1992). As much as 125 nt could be inserted at nt 490 without effect on IRES function, whereas deletion of only

---

Legend to Figs. 1–3

Fig. 1. Consensus secondary structure of the Cardiovirus IRES, based on the sequence of encephalomyocarditis virus (EMCV) strain R and the secondary structures proposed by DUKE et al. (1992) and PILIPENKO et al. (1989). Domains D-L are named alphabetically as proposed by Duke et al. (1992). Nucleotides and AUG triplets (indicated by *thick lines*) in the EMCV 5' NTR are *numbered*, and the initiation codon AUG_{834} is *boxed*. *Shaded blocks* indicate nucleotides that are conserved in 18 different cardiovirus strains. (For references see Sect. 3)

Fig. 2. Secondary structure of the *Aphthovirus* IRES, based on the sequence of foot and mouth disease virus (*FMDV*) strain O_1K (FORSS et al. 1984) and on the secondary structure proposed by PILIPENKO et al. (1989). Nucleotides in the FMDV 5' NTR are *numbered*, the initiation codon is *boxed*, and domains D-L are named alphabetically in accordance with the nomenclature for EMCV (DUKE et al. 1992). *Shaded blocks* indicate nucleotides that are conserved in 17 different FMDV strains and serotypes. (For references see Sect. 3)

Fig. 3. Secondary structure of the echovirus (EV) type 22 IRES, based on the sequence of the Harris strain of EV22 (HYPPIA et al. 1992), and the secondary structures proposed for *Cardiovirus* and *Aphthovirus* IRES elements (DUKE et al. 1992; PILIPENKO et al 1989). Nucleotide positions in the EV22 5' NTR are *numbered*, the initiation codon is *boxed*, and domains D-L are named alphabetically as proposed for the *Cardiovirus* IRES (DUKE et al. 1992). Nucleotides that are conserved in EV22 (HARRIS) and encephalomyocarditis virus (EMCV) (R) are *shaded*

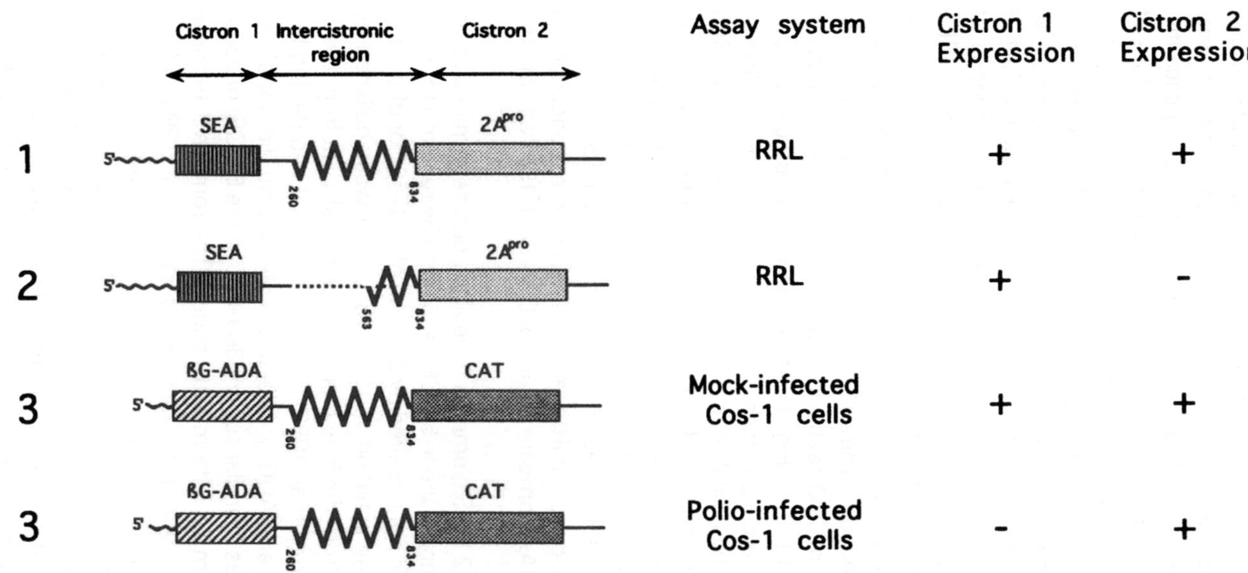

Fig. 4. Structure and expression of dicistronic mRNAs used for in vitro and in vivo translation. The 5'NCRs of poliovirus (*1,2*) and β-globin (*3*) are shown as *thin zig-zag lines,* and coding regions are *boxed* and correspond to the reporter genes SEA (viral oncogene *sea*), $2A^{pro}$ (poliovirus 2A proteinase), βG-ADA (chimæric β-globin/human adenosine deaminase polypeptide), and CAT (bacterial chloramphenicol acetyltransferase) as indicated. Segments of the encephalomyocardits virus (EMCV) 5'NTR in the intercistronic space are drawn as *thick zig–zag lines.* RNAs 1 and 2 were transcribed in vitro and translated in rabbit reticulocyte lysate (*RRL*), whereas RNA 3 was transcribed in vivo, and translated in COS-1 cells 4 h after either mock infection or poliovirus infection. Expression of cistrons 1 and 2 is an indicated. Adapted from previous reports (JANG et al. 1988, 1989)

three nt at nt 560 abolished translation (WITHERELL and WIMMER 1993). Third, mutations within H, J and K domains of the IRES severely impair initiation (JANG and WIMMER 1990; OUDSHOORN et al. 1990; SCHEPER et al. 1991). Significantly, additional compensatory substitutions that restore the secondary structure of domain H also restored IRES function (JANG and WIMMER 1990). The fourth line of evidence comes from deletion analysis (BOROVJAGIN et al. 1991; DUKE et al. 1992; EVSTAFIEVA et al. 1991; JANG 1989; JANG et al. 1988, 1989; JANG and WIMMER 1990; KAMINSKI et al. 1990). Nucleotide substitution and serial deletion of the EMCV 5'NCR from its 3' border showed that the number, but not the sequence of nucleotides in the region upstream of AUG_{834} are critical for wild-type IRES activity: deletion of 9 nt reduced initiation fivefold, whereas substitution of these residues had no effect. However, IRES function was almost totally abrogated by deletion of two additional nucleotides. The EMCV IRES does not have a precise 5' border: serial deletions from nt 280 result in a progressive loss of function. Deletion of domains D, E and F (see Fig. 1) halves translation efficiency, and further deletion of domain G reduces translation to 20% of wild-type levels. Translation is reduced further by partial or total deletion of domain H, and IRES function is abolished by deletion past nt 500. Domains H, I, J and K thus appear to be critical for IRES function, whereas domains D, E, F, and G appear to supplement its activity. Consistent with this conclusion, all internal deletions within these four core domains abrogated or severely impaired IRES function. In general, the effect of an alteration to the IRES appears to be more pronounced in the context of dicistronic mRNAs (compare results of DUKE et al. 1992 and of JANG and WIMMER 1990).

The borders of other type 2 IRES elements have also been determined using deletion analysis; larger segment of these 5'NCRs appear to be required for IRES function. TMEV translation was not impaired by deletion of nt 1–500 or of the ten nt immediately upstream of the initiation codon AUG_{1065}, but was abolished by deletion of an additional 94 nt from the 5' end or of an additional 16 nt from the 3' end of the 5'NCR (BANDYOPADHYAY et al. 1992; HUNT et al. 1993). Lengthening the sequence between domain L and AUG_{1065} (albeit by only two nt) did not affect translation. The FMDV IRES has a similar size (nt 369–805 of the 5'NCR, or about 435 nt) and the loss of 30 nt from the 5' terminal or about 50 nt from the 3' terminal abolished IRES function (BELSHAM and BRANGWYN 1990; KUHN et al. 1990). Mutations throughout this region indicated that domains H, I, J and L are most important for FMDV IRES function. There are some discrepancies between reports concerning the borders of the HAV IRES (BROWN et al. 1991, 1994; GLASS et al. 1993), but it appears that almost the entire 5'NCR downstream of the unstructured pyrimidine-rich tract (i.e., nt 150–734) is required for efficient translation, although further truncation at the 5' end may leave some residual activity.

These observations indicate that the structure of IRES elements is critical for their ability to promote initiation, in clear distinction to the lack of important structural features in cellular mRNAs (such as β-globin) which are translated by a cap-dependent mechanism (KOZAK 1994). They also provide an attractive explanation for the structural conservation of IRES elements in EMCV, EV22, FMDV, and HAV.

6 Cis-Acting Elements Within Type 2 Internal Ribosome Entry Sites

Most sequence variation in type 2 IRES elements takes the form of compensatory substitutions that serve to maintain IRES structure. The implication that IRES structure is important for function has been confirmed; as noted above, small deletions or insertions in EMCV, HAV and FMDV IRES elements severely impaired their function (Borovjagin et al. 1991; Duke et al. 1992; Evstafieva et al. 1991; Glass et al. 1993; Kühn et al. 1990). However, the conservation of specific sequence motifs suggests that type 2 IRES elements also contain cis-acting RNA elements. A number of potential structures of this type have been identified, but only three have been characterized in detail.

The importance of EMCV domain H in IRES function was revealed by deletion analysis (Borovjagin et al. 1991; Jang and Wimmer 1990). This domain consists of a five nt loop, and an internal eight nt bulge separating two short helices (Fig. 1). Deletion of the upper stem or disruption of its base pairing by substitution impaired initiation severely; significantly, compensatory substitutions which restored the structure of this stem also restored translational efficiency (Borovjagin et al. 1991; Jang and Wimmer 1990). A three nt insertion into domain H of TMEV severely impaired IRES function in vivo and attenuated neurovirulence, but curiously had no effect on TMEV translation in RRL (Bandyopadhyay et al. 1993). The trans-acting factor p57/PTB (see Sect. 8) binds specifically to EMCV RNA transcripts consisting of domain H with or without flanking residues; this interaction was abolished both by the deletion and by the substitutions in it that were described above, but binding was restored by the same compensatory substitutions that restored translational activity (Borovjagin et al. 1991; Hellen et al. 1993; Jang and Wimmer 1990; Witherell et al. 1993). A probe corresponding to just the loop and upper helix of this domain did not bind p57. Two additional p57 binding sites within the EMCV 5'NCR (encompassed by nt 315–377 and nt 740–837) have been identified, but they have not yet been characterized in detail (Borovjagin et al. 1991; Witherell et al. 1993; Witherell and Wimmer 1994). The structure and sequence of domain H in the FMDV IRES resembles that of the corresponding domain in EMCV, and it also binds p57/PTB specifically (Luz and Beck 1990, 1991). It plays an important role in FMDV translation, since deletion of three nt from the internal bulge halved the efficiency of FMDV translation (Kühn et al. 1990). N[U → C] substitution within the upstream p57 binding site in the 5'NCR of an FMDV isolate from persistently infected cells enhanced translation in vivo, although the basis for this effect is not known (Martinez-Salas et al. 1993). Binding of p57 to a second, downstream site in the FMDV IRES (nt 1227–1287 in Fig. 2) was reduced by substitutions within a pyrimidine-rich tract downstream of domain L that also impaired translation (Luz and Beck 1991).

EMCV RNA transcripts corresponding to domain H competitively inhibit binding of p57 to the poliovirus IRES, which has a lower affinity for this factor, and can inhibit IRES-dependent translation of poliovirus RNA but not cap-dependent translation of β-globin (LUZ and BECK 1991; PESTOVA et al. 1991). However, RNA transcripts corresponding to domains J and K are able to inhibit cap-dependent translation, an observation suggesting that these domains may interact with factors involved in both initiation mechanisms (DUKE et al. 1992; EVSTAFIEVA et al. 1991), which may include eIF2/2B (SCHEPER et al. 1991). Mutations in these domains impair IRES-dependent translation of EMCV RNA, and the ability of IRES-specific RNA transcripts to act as competitive inhibitors of translation (DUKE et al. 1992; OUDSHOORN et al. 1990; SCHEPER et al. 1991).

A pyrimidine-rich tract (Y*n*, where $n = 6$–9 nt) upstream of picornavirus initiation codons was first noted in the FMDV serotypes C_1 and O_1K (BECK et al. 1983). These two motifs separated by a random spacer X*m* (where $m = 15$–20 nt) were soon recognized as a characteristic of all aphthoviruses and cardioviruses (CLARKE et al. 1987; FORSS et al. 1984; OHARA et al. 1988). All FMDV isolates except SAT-3 also have a Yn-Xm motif upstream of the second initiation codon (CLARKE et al. 1987). A similar Yn-Xm-AUG motif occurs at the 3' end of the IRES in all other picornavirus 5'NCRs (Fig. 5; JANG et al. 1990). The 3' borders of the IRES and the 5'NCR coincide in cardioviruses, hepatoviruses and presumably in EV22, so the AUG triplet of the Yn-Xm-AUG motif is the initiation codon, whereas in enteroviruses and rhinoviruses it is inactive in translation. Sequence similarity in this region extends downstream of the AUG triplet (Fig. 5; HUNT et al. 1993; PESTOVA et al. 1994), even though these residues are noncoding in type 1 IRES elements and part of the viral polyprotein downstream of type 2 IRES elements.

The influence of all four components of this motif on initiation directed by type 2 IRES elements have been investigated experimentally. Deletion analysis

Virus	Nt.	Yn	Xm	AUG - ACAAYYA
PV1M	556	GUGUUUCCUUUU.......	15	CUUAUGGUGACAAUCACAGA
HRV2	547	GUGUUUCACUUU......	14	CUUAUGGUGACAAUAUAUAC
EMCV-RR	798	GGUUUUCCUUUG.......	14	UAUAUGGCCACAACCAUGGA
FMDV A12	>1000	GCACCUUUUUCU.......	14	UUUAUGAACACAACCAACUG
EV22	706	UGGUUUCCUUUU.......	13	AUUAUGGAGACAAUUAAGAG
HAV-LA	710	GUUUUUCCUCAU.......	9	AUAAUGAAUAUGUCCAAACA
HCV-H	329			AUCAUGAGCACGAAUCCUAA
Consensus				xUUAUGRxxACAAYYA

Fig. 5. Alignment of nucleotide sequences at the 3' border of the IRES elements of representative members of the five picornavirus genera *Enterovirus*: poliovirus type 1 (Mahoney); PV1M, *Rhinovirus* (human rhinovirus type 2; HRV2), *Cardiovirus* (encephalomyocarditis virus Rueckert strain; EMCV-R), *Aphthovirus* (foot-and-mouth disease virus type A12;FMDV A12) and *Hepatovirus* (hepatitis A virus Los Angeles strain; HAV LA), of echovirus 22 (EV22) and of hepatitis C virus (HCV). The HCV and EV22 sequences are as described (TSUKIYAMA-KOHARA et al. 1992 and HYPPIA et al. 1992, respectively); references to all other picornavirus nucleotide sequences are given in a review (STANWAY 1990). The alignment of pyrimidine residues (*Yn*) and the conserved AUG triplet of the Yn-Xm-AUG motif and the downstream ACAAYYA motif have been noted previously (HUNT et al. 1993; JANG et al. 1990; NICHOLSON et al. 1991; PESTOVA et al. 1991, 1994)

first implicated the Yn tract as a potential cis-acting element in the EMCV IRES (JANG and WIMMER 1990);substitutions within this tract reduced FMDV translation up to 20-fold (KÜHN et al. 1990). The Yn tract has been considered as part of a tertiary interaction with the A-rich bulge between J and K domains (LE et al. 1993) or the site of a quaternary interaction with either the 3' end of 18S rRNA (BECK et al. 1983) or with a regulatory protein such as p57 (KÜHN et al. 1990). None of these possibilities appears compatible with the observation that substitution of the entire Yn tract by A residues had only a modest effect on EMCV translation (KAMINSKI et al. 1994). The Xm spacer separating Yn and AUG elements is the least conserved region in all IRES elements, although its length is constant. Substitutions within it have no effect on initiation efficiency (DAVIES and KAUFMAN 1992) and it may simply act as a relatively unstructured segment which maintains the initiation codon at an appropriate distance from the IRES (PILIPENKO et al. 1994).

Some cardiovirus and aphthovirus isolates contain AUG triplets in the spacer (ESCARMIS et al. 1992; PALMENBERG et al. 1984; SANGER et al. 1987) and UGAROVA noted that one such EMCV triplet (AUG_{826}) is not active in translation eventhough its context (ACG*AUG*A) is more favourable for initiation than the context (UAU*AUG*A) of the initiation codon (AUG_{834}) of the viral polyprotein (UGAROVA 1987). Similar observations probably apply to the FMDV serotype A24 (SANGAR et al. 1987) and suggest that initiation promoted by type 2 IRES elements involves entry of initiation-competent ribosomes at the initiation codon without scanning upstream sequences. Initiation at AUG_{826} occurred in preference to initiation at AUG_{834} on translation of mRNAs containing only the last 96 nt of the EMCV 5' NCR (KAMINSKI et al. 1990). The upstream triplet is thus inherently capable of initiating translation in the absence of the IRES, confirming the hypothesis that the IRES causes initiating ribosomes to bypass AUG_{826} and to bind directly at AUG_{834}. This observation suggests that there may be IRES-specific determinants of initiation codon recognition.

A reduction in the length of the Xm spacer in EMCV reduced initiation at AUG_{834} and led to preferential initiation at AUG_{846},four triplets downstream of AUG_{834} (JANG and WIMMER, unpublished results). By contrast, an insertion of eight nt in the spacer upstream of AUG_{826} did not switch initiation from AUG_{834} to AUG_{826} Kaminski et al. 1994. A minimum separation of AUG_{834} from upstream IRES elements is thus necessary for initiation at this codon, but there must be additional determinants that promote its recognition. The local context is a second determinant, since an A → U substitution at the –3 position impaired initiation at AUG_{834} and, curiously, resulted in initiation at three downstream codons (DAVIES and KAUFMAN 1992). The sequences downstream of AUG_{834} appear to be a third determinant: residues downstream of the AUG triplet at the 3' border of all IRES elements have a similar sequence (Fig. 5) and their substitution in EMCV impaired the efficiency of initiation (DAVIES and KAUFMAN 1992; HUNT et al. 1993).

7 Trans-Acting Factors Involved in Internal Ribosome Entry Site-Dependent Translation

The EMCV IRES is active in a broad range of mammalian, avian and amphibian cells, but it is not translated in wheat germ lysates which are able to support translation of β-globin mRNA (JEN and THACH 1982; unpublished data). Translation of EMCV RNA by internal initiation may therefore require novel activities that are not required for cap-dependent translation. Conversely, some factors required for cap-dependent initiation of translation may not be required for IRES function, since EMCV RNA is translated in poliovirus-infected cells, in which eIF-4γ is cleaved and its interaction with eIF-3 is altered (ETCHISON et al. 1982; ETCHISON and SMITH 1990). However, some characteristics of the inhibition of host cell protein synthesis in Cardiovirus-infected cells are consistent with Cardiovirus IRES elements possessing a higher affinity than host mRNAs for a limiting factor that is involved in both mechanisms.

Cap-dependent initiation of eukaryotic protein synthesis is a complex process in which ribosomal sununits bind initiator tRNA, attach to an mRNA template and translocate to the first downstream AUG triplet that has a favourable sequence context for initiation, thereby aligning this codon with the anticodon of the intiator tRNA, and assemble to form an initiation complex. A pathway describing the sequential use of initiation factors has been proposed to account for this process, although biochemical details of some steps remain obscure (reviewed by HERSHEY 1991; KOZAK 1991; MERRICK 1992; TRACHSEL 1991). Reports concerning the role of canonical initiation factors in Cardiovirus translation are not wholly consistent, probably because different cell types and protocols were used in the preparation of fractionated protein synthesizing systems, and because purification of some factors may not have been sufficiently rigorous to exclude contamination. The factor eIF-4A was first identified because of its specific requirement for EMCV RNA translation (WIGLE 1973; WIGLE and SMITH 1973) and was subsequently also found to stimulate translation of mengovirus and FMDV RNAs much more than translation of (capped) globin mRNA (BLAIR et al. 1977). The involvement of eIF-4A in EMCV translation is supported by the impairment of translation by dominant negative mutants of eIF-4A in RRL (PAUSE et al. 1994). However, the RNA binding activity of eIF-4A is low, sequence nonspecific and strictly ATP-dependent (ABRAMSON et al. 1987), and its involvement in IRES function is therefore puzzling in light of the ATP-independence of preinitiation complex formation on EMCV RNA (JACKSON 1991b). eIF-4B greatly enhances the affinity of eIF-4A for RNA (ABRAMSON et al. 1988) and it has also been implicated in EMCV translation (GOLINI et al. 1976; BAILGONI et al. 1978), although these claims should be treated cautiously (SONENBERG 1987). Nevertheless, stimulation of IRES-dependent initiation by the ssRNA binding factors eIF-4A, eIF-4B, eIF-4E and eIF-4F has recently been demonstrated using artificial bicistronic mRNAs and wild-type EMCV mRNA (ANTHONY and MERRICK 1991; PAUSE et al. 1994; SCHEPER et al. 1992). The involvement of eIF-4E and eIF-4F is surprising because earlier reports had indicated that these factors did not stimulate translation of EMCV and mengovirus mRNAs (Daniels- McQueen et al. 1983; Edery et al. 1984; Sonenberg et al. 1980). The activity of eIF-2 is regulated by phosphorylation (Jackson 1990) and it is therefore a plausible candidate for a limiting factor whose preferential binding to Cardiovirus RNA would inhibit host cell protein synthesis. Filter-binding assays have shown that mengovirus has a 30-fold higher affinity for eIF-2 than globin mRNA (Rosen et al. 1982), and nuclease protection assays have shown that eIF-2 binding sites on mengovirus mRNA coincide with ribosome binding sites (Perez-Bercoff and Kaempfer 1982). Domains J and K may constitute a high affinity eIF-2/2B binding site within the EMCV IRES (Scheper et al. 1991). However, no stimulation of EMCV translation was observed in RRL on addition of either eIF-2 or eIF-2B, indicating that the concentration of these factors was not limiting (Svitkin et al. 1994).

An alternative approach that has been used to identify factors that interact with the EMCV IRES is UV cross-linking of cellular polypeptides present in cell-free lysates to [^{32}P-UTP]-labeled IRES-specific RNA transcripts. A large number of RNA-binding proteins become labeled in this assay: the principal moieties are termed p36, p43, p50, p52, p57/58, p69 and p110, corresponding to their molecular weights(Borovjagin et al. 1990; Jang and Wimmer 1990; Pestova et al. 1991; Witherell and Wimmer 1994). The HAV IRES binds to four major polypeptides: p30, p39, p57 and p110 (Chang et al. 1993). p57/58 refers collectively to polypeptides that are resolved by SDS-polyacrylamide gel electrophoresis as a set of two to four closely migrating bands: they are the most prominently labeled polypeptides after UV cross-linking to the EMCV IRES and also interact strongly with the FMDV IRES (Luz and Beck 1990, 1991). Two of the polypeptides listed above have been identified: p57 is the pyrimidine tract-binding protein (PTB), a member of the heterogenous nuclear ribonucleoprotein family (Borman et al. 1993; Hellen et al. 1993,1994). p52 is the La autoantigen, which is involved in RNA polymerase III transcription termination and also has a predominantly nuclear localization (Meerovitch et al. 1993). p52 was initially identified by its ability to bind to a segment of the poliovirus 5'NCR (Meerovitch et al. 1989); this factor can also be immunoprecipitated from the set of cytoplasmic polypeptides that become labeled after UV cross-linking to the EMCV IRES (Fig. 6). However, no role for this polypeptide in EMCV translation has been reported.

8 The Role of p57/PTB in Internal Ribosome Entry Site-Dependent Initiation of Translation

There is considerable evidence to suggest that p57 is involved in initiation of picornavirus translation directed by both type 1 and type 2 IRES elements. The EMCV IRES contains three p57/PTB binding sites corresponding to domains E/F,

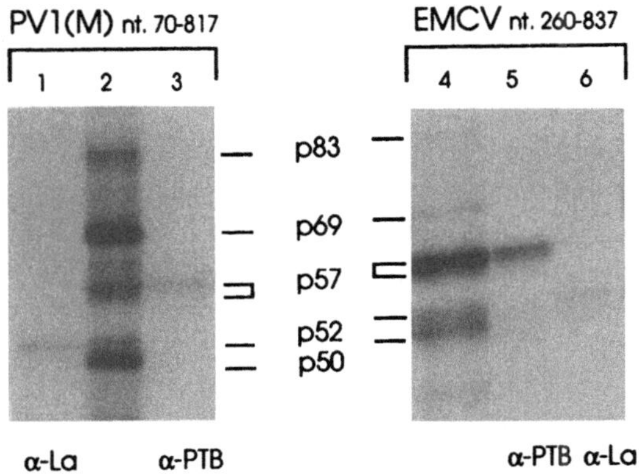

Fig. 6. Immunoprecipitation of p57 by anti-PTB antibodies (lanes *3* and *5*) and of cross-linked p52 by anti-La antibodies (lanes *1* and *6*) after UV cross-linking of polypeptides in a translation competent HeLa cell-free extract to [^{32}p-UTP]-labeled RNA transcripts corresponding to poliovirus 1(M) (nt 70-817; lanes *1-3*) or to EMCV (nt 260-837; lanes 4-6). The crosslinked and immunoprecipitated polypeptides were resolved by electrophoresis in an SDS/10%–20% polyacrylamide gradient gel. The positions of cross-linked poly-peptides are indicated between the two panels

domain H and nt 740–837; binding sites corresponding to the latter two domains have been identified in the FMDV IRES (Luz and Beck 1991). Cross-linking of p57/PTB to the central EMCV binding site was abolished by deletion of the 5' half of domain H, or disruption of its structure by nucleotide substitution, and was restored by compensatory mutations that restored the structure and funcrtion of this domain (Jang and Wimmer 1990). There is thus a strong correlation between the structure of the EMCV IRES, the efficiency with which it promotes translation, and the strength of its interaction with p57/PTB. A similar correlation has been noted for the FMDV IRES (Luz and Beck 1991). A role for p57/PTB in type 2 IRES function is further supported by the observation that immunodepletion of this factor from a HeLa cell free lysate impaired EMCV translation but did not affect translation of β-globin mRNA (Hellen et al. 1993). The formation of 48S pre-initiation complexes on EMCV mRNA in vitro was inhibited by sequestration of endogenous p57 using EMCV IRES-specific transcripts, but was rescued by addition of recombinant PTB-1 (Borovjagin et al. 1994). Four observations suggest that p57/PTB is also involved in type 1 IRES function: (1) it is UV cross-linked to poliovirus and *Rhinovirus* IRES elements (Borman et al. 1993; Hellen et al. 1994; Luz and Beck 1991; Pestova et al. 1991); (2) poliovirus translation and binding of p57 to its IRES are competitively inhibited by RNA transcripts corresponding to domain H of the EMCV IRES (Pestova et al. 1991); (3) immunodepletion of p57/PTB impaired poliovirus translation (Hellen et al. 1993); and (4) *Rhinovirus* translation in RRLs is stimulated greatly by HeLa cytoplasmic fractions containing p57/PTB (Borman et al. 1993). p57 is detected in higher concentration in HeLa, BHK, and Krebs II cell extracts than in RRL and appears to be present mainly in the ribosomal salt wash, with only trace amounts apparent in the postribosomal fraction (BOROVJAGIN et al. 1990; HELLEN et al. 1994; JANG and WIMMER 1990; LUZ and BECK 1991).

PTB, also known as hnRNP I, is a predominantly nuclear protein (GHETTI et al. 1992). It was initially identified by UV cross-linking to the pyrimidine tract upstream of the 3' splice site in pre-mRNAs (GARCIA-BLANCO et al. 1989) but it can also bind other pre-mRNA elements that do not all contain long pyrimidine tracts (BENNETT et al. 1922b; MULLIGAN et al. 1992). Its binding is ATP-independent (GARCIA-BLANCO et al. 1989; MULLEN et al. 1991) and optimal at low Mg^{2+} and K^{+} concentrations (MICHAUD and REED 1991; NORTON and HYNES 1993). PTB is capable of self-association, and interacts specifically with a 100 kDa PTB-associated splicing factor (PSF); it copurifies with PSF and a 33–35 kDa polypeptide which has been identified as hnRNP A1 (BOTHWELL et al. 1991; PATTON et al. 1993). Anti-PTB serum can block splicing at an early stage, probably before formation of the first "commitment" complex (MOORE et al. 1993), but the role of PTB in splicing has not been precisely defined. It is not a major constituent of "committed" pre-spliceosomal complexes nor is it required for their formation, but it can nevertheless be detected in spliceosomes by immunoprecipitation (BENNETT et al. 1992a, b; JAMISON et al. 1992; PATTON et al. 1991). PTB associates readily with specific pre-mRNAs in splicing extracts, probably in competition with other polypeptides such as hnRNP C, but it is also readily displaced by salt treatment and is released from in vivo isolated hnRNP complexes by nuclease digestion before other hnRNPs (BENNETT et al. 1992a; GHETTI et al. 1992). The role of PTB may therefore be to interact with pre-mRNAs at an early stage in spliceosome assembly and then to exchange with another factor, such as U2AF (BENNETT et al. 1992a, b). PTB could thus determine splice-site selection by influencing the site of binding of this or other factors to pre-mRNAs (e.g., MULLEN et al. 1991). However, a negative role for PTB in splicing has also been suggested (MULLIGAN et al. 1992; NORTON and HYNES 1993).

cDNAs of four human PTB isoforms and murine and rat homologs have been sequenced, revealing that PTB isoforms contain 528–557 amino acid residues and are generated by alternative splicing (BOTHWELL et al. 1991; BRUNEL et al. 1991; GHETTI et al. 1992; GIL et al. 1991; PATTON et al. 1991). PTB contains four repeated domains of 80 amino acids which can be aligned with RNA recognition motif (RRM) consensus sequences found in other RNA binding proteins (KENAN et al. 1991). The RRM domain contains a highly conserved octamer (RNP-1) sequence separated from a hexamer (RNP-2) sequence by 30–35 amino acid residues, and although PTB lacks these canonical sequences, the pattern of conserved hydrophilic residues in each domain suggests that they adopt a similar structure. The two NH_2-terminal RRM domains in murine PTB are sufficient for specific binding to the pyrimidine tract in pre-mRNAs (BOTHWELL et al. 1991).

Binding of PTB to IRES elements is also ATP-independent and optimal at low K^{+} concentrations; however, ionic interactions contribute less than 40% of the total free energy on formation of PTB-RNA complexes (BOROVJAGIN et al. 1990; LUZ and BECK 1991; WITHERELL et al. 1993). The remainder may come from conformational changes on binding or from hydrophobic interactions between

PTB molecules. PTB is capable of self-association, but the stoichiometry of its interaction with the IRES has not yet been determined. Proteins containing multiple RRM domains can interact simultaneously with several RNA sequences and individual RRM domains can have distinct RNA binding specificities (BURD et al. 1991; NIETFELD et al. 1990). PTB could bind to more than one of the binding sites within an IRES at a time since it contains four RRM domains; moreover, these binding sites could have dissimilar sequences and/or structures.

The functional interaction of p57/PTB with an IRES element is likely to depend on additional factors. This conclusion is based on two observations; first, the binding specificities of both PTB from nuclear extracts and of p57 from ribosomal salt wash fractions were reduced on purification, but could be restored by addition of HeLa S10 cytoplasmic extracts to recombinant PTB (GIL et al. 1991; HELLEN et al. 1993,1994; WITHERELL et al. 1993). Second, stimulation of type 1 IRES-dependent initiation in RRL by a 97 kDa polypeptide is dependent on the presence of p57/PTB, which itself has little stimulatory activity (BORMAN et al. 1993). However, p97 appears to be antigenically distinct from PSF, the 100 kDa polypeptide associated with PTB in splicing extracts (T. Pestova, personal communication).

The molecular basis for the involvement of PTB in picornavirus translation has not yet been established; it may act as a nucleation site for general initiation factors or even constitute the recognition site for ribosome binding. Alternatively, it could modulate the structure of the IRES so that factors or ribosomes are able to bind to it directly.

9 A Comparison Between Type 1 and Type 2 Internal Ribosome Entry Sites

Picornavirus IRES elements have been divided into two major groups on the basis of functional characteristics and structural similarities (JACKSON et al. 1990; JANG et al. 1990). Although sequence conservation between members of each of the two groups is extensive and quite strong (Sect. 3; LE and ZUKER 1990; RIVERA et al. 1988), the only conservation that has been noted between the two groups is the pyrimidine-rich tract within the Yn Xm-AUG motif (JANG et al. 1990; NICHOLSON et al. 1991; PESTOVA et al. 1991). Reexamination of the sequences of poliovirus (PV) and EMCV 5'NCRs reveals that there are additional sequence similarities between these representative type 1 and type 2 IRES elements (Fig. 7). The region of sequence conservation corresponds to nucleotides between domain G and the initiation codon region in EMCV, and to nucleotides encompassed by domains II–VI of the PV 5'NCR. Sequence conservation therefore extends beyond the 3' borders of the two IRES elements (Fig. 5), which is notable since these residues are noncoding within type 1 IRES elements but constitute part of the polyprotein coding region downstream of type 2 IRES elements. Initiation occurs about 30 nt downstream of the Rhinovirus IRES and up to 154 nt downstream of Enterovirus IRES elements (BORMAN and JACKSON 1992; DORNER et al. 1982; PESTOVA et al. 1994). Nucleotides within both IRES elements are base-paired rto form a series of hairpins (see Sect. 4 and Figs. 1 and 7). Some domains within these two elements appear to be structurally related and to have similar relative positions. For example, the domains EMCV-H and PV-III have a similar length, contain internal bulges and are preceded by short hairpins, whereas EMCV-I and PV-V both contain long central stems whose distal third is interrupted by three or four internal hairpins. The PV IRES appears to lack structures equivalent to EMCV domains K and L, and there are many smaller structural differences between it and the EMCV IRES, but the level of sequence conservation and structural similarity suggests that the tertiary structures of these two elements may resemble one another.

The two IRES groups function most efficiently in vitro under different conditions, specifically with respect to ATP dependence and optimum ionic and RNA concentrations (Sect. 2; JACKSON 1989, 1991a, b; PESTOVA et al. 1994; VILLA-KOMAROFF et al. 1975). Type 1 IRES-dependent initiation is inefficient in some cells and cell-free extracts (such as Xenopus oocytes and RRL) in which type 2 IRES elements function efficiently (BORMAN and JACKSON 1992; DORNER et al. 1984; LASKEY et al. 1972; PELLETIER et al. 1988), but is stimulated by supplementation of RRL with protein factors from HeLa or Krebs II ascites cells (DORNER et al. 1984; SVITKIN et al. 1988), indicating that RRL is deficient in a factor that is required for type 1 IRES function. Initiation of PV translation occurs downstream of the IRES (whereas the initiation codon coincides with the 3' border of type 2 IRES elements) but this deficiency appears not to correspond to factors required for ribosome scanning (SVITKIN et al. 1994). The initiation codon for the PV polyprotein can be placed at the 3' border of the IRES by deletion of the intervening 154 nt spacer (PESTOVA et al. 1994), but translation in RRL is still inefficient, so the stimulatory activity present in HeLa and Krebs cells appears to be required for internal entry per se rather than for initiation events subsequent to ribosome binding. The 52 kDa La polypeptide (MEEROVITCH et al. 1993; SVITKIN et al. 1994) and a high Mr (>300 00 kDa) complex which contains p57 and p97 (SVITKIN et al. 1988; BORMAN et al. 1993) are candidates for the stimulatory activity, and are discussed in detail elsewhere in this volume. Several RNA binding proteins can be UV cross-linked to type 1 and to type 2 IRES elements, including p52, p57 and p110 (Sect. 7). The roles (if any) of p52 and p110 in EMCV translation are not known, but a common requirement for p57 by both types of IRES element (Sect. 8) indicates that they contain functionally related cis-acting elements and may be mechanistically related.

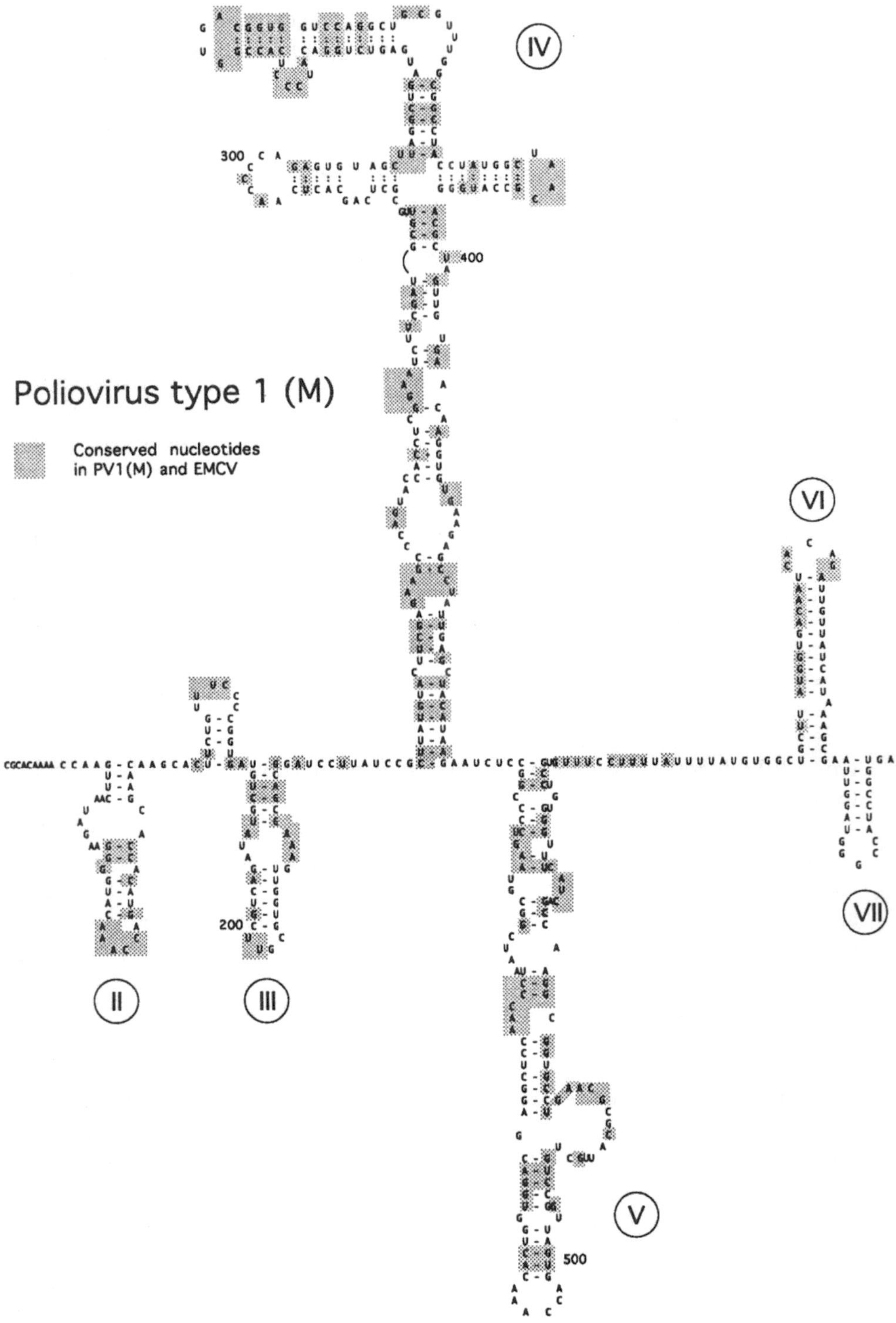
Poliovirus type 1 (M)
Conserved nucleotides in PV1(M) and EMCV
IV
300
400
VI
VII
II
III
200
V
500

10 Applications of the Encephalomyocarditis Virus Internal Ribosome Entry Site

IRES elements function in the absence of viral gene products and additional viral sequences and can therefore be used in heterologous contexts. The EMCV IRES promotes efficient initiation in a particularly broad range of eukaryotic cells. It has therefore been used both to simply enhance expression levels and, specifically, to exploit the internal ribosome entry mechanism. A variety of expression systems utilize bacteriophage T3 or T7 promotors and thier respective RNA polymerases to achieve high levels of expression in mammalian cells. Transcripts synthesized by these polymerases are not capped and may contain structural elements (hairpins) inhibitory to capping or scanning, but inclusion of an IRES element downstream of the promoter yields RNA that is translated efficiently in variety of cell types by avoiding the limitations of 5' end-dependent initiation. The first application of the EMCV IRES to enhance expression at the level of translation was reported by Elroy-Stein et al. (1989), who incorporated the element into the vaccinia virus expression system, and several related publications followed (Deng et al. 1991; Elroy-Stein et al. (1989; Elroy-Stein and Moss 1990; Mirochnitchenko et al. 1994; Vennema et al. 1991; Zhou et al. 1990). The EMCV IRES was identified by translation of bicistronic mRNAs (Jang et al. 1988, 1989), and the same principle has been exploited in constructing bicistronic vectors for high level expression of foreign genes (Adam et al. 1991; Ghattas et al. 1991; Morgan et al. 1992; Kaufman et al. 1991). One cistron in such an mRNA typically encodes a reporter gene (such as chloramphenicol acetyltransferase) and the second contains a selectable marker (such as neomycin phosphotransferase). The presence of a single transcription unit avoids potential problems of promoter suppression, and the presence of an IRES reduces the probability of deletion of the reporter gene. This strategy has obvious potential applications in gene therapy, particularly since multiple protein subunits or heterologous polypeptides can be expressed (Jang et al. 1989). The EMCV IRES does not contain fortuitous splice sites (and is thus not spliced after transcription in vivo), so integration into host DNA of vector sequences containing the IRES fused to either a reporter gene or a selectable marker can result in transcription from endogenous promoters and can consequently be used for the selection of specific recombinants (Wood et al. 1991) or the detection of (developmentally regulated) transcription (Kim et al. 1992).

The picornavirus genome is monocistronic and encodes a single polyprotein which is proteolytically processed to yield a large number of distinct proteins. Insertion of an IRES into PV RNA generates a dicistronic genome which should be

◄

Fig. 7. Sequence and structural similarities between representative type 1 and type 2 IRES elements. The secondary structure of the (type 1) IRES of poliovirus type 1 (Mahoney) has been drawn to emphasize similarities with the (type2) EMCV IRES; conserved nucleotides in these two IRES elements were identified by sequence and structural comparison and are indicated by *shading*. Nucleotides in the PV1(M)5'-NTR are *numbered*, and domains I–VII are named as described by Hellen et al. (1994)

a viable self-replicating entity, provided that the insertion does not disrupt the coding sequence of an essential gene product and does not interfere with proteolytic processing of the polyprotein. Three insertion sites have been identified that meet these criteria and have yielded viable dicistronic polioviruses (Fig. 8). The scissile bond between P1 (capsid protein) and P2-P3 (nonstructural protein) regions was chosen as the first insertion site because this bond is cleaved cotranslationally, and the P1 partial polyprotein is subsequently cleaved in trans. The resulting virus W1-P1/E/P2,3-1 was genetically stable but its replication was impaired; partial deletion of the EMCV IRES abolished viral replication (MOLLA et al. 1992). These results confirm the existence of the EMCV IRES and provide a novel strategy for genetic dissection of the PV polyprotein. Insertion of the EMCV IRES between $2A^{pro}$ and 2B yielded a viable virus W1-P1, 2A/E/2BC, P3-1 with impaired replication characteristics, indicating that $2A^{pro}$ can function in trans in all events subsequent to its separation from P1 (MOLLA et al. 1993). Dicistronic genomes lacking $2A^{pro}$ failed to replicate, as did genomes in which $2A^{pro}$ had been modified by partial deletion or active site substitution, suggesting a role for this polypeptide in genome replication. Insertion of the EMCV IRES into all other cleavage sites of the polyprotein abolished viral replication, possibly due to aberrant proteolytic processing of the P2-P3 region and/or disruption of an active precursor (e.g., 2BC, 3AB or 3CD; WIMMER et al. 1993). However, insertion of the EMCV IRES at nt 630 in the PV 5'NCR produced the viable, genetically stable virus W1-PNENPO which has two different IRES elements arranged in tandem (Fig. 8); replacement of the PV IRES with the EMCV IRES yielded the virus W1-P108ENPO in which translation is dependent solely on the heterologous EMCV IRES (ALEXANDER et al. 1994).

To investigate the potential of dicistronic PVs as expression vectors, the coding sequences for chloramphenicol, acetyltransferase and luciferase were inserted between the tandem IRES elements in W1-PNENPO, yielding the dicistronic PV genomes pDICAT and pDILUC. Both RNAs were replication competent, but only the former RNA was encapsidated, yielding the genetically unstable virus W1-DICAT (ALEXANDER et al. 1994). The pDICAT and pDILUC genomes are 17% and 31% longer than the 7500 nt long wild-type RNA, respectively, so 7800 nt, may correspond to the upper packaging limit of PV virions. Other coding sequences could obviously be substituted for these reporter genes; expression of heterologous coding sequences might be used to elicit antigenic responses, whereas duplicated PV genes could be used to complement genetic defects in mutant PV genomes (WIMMER et al. 1993).

11 Summary

Picornavirus 5' NCRs contain IRES elements that have been divided into two groups, exemplified by PV (type 1) and EMCV (type 2).These elements are

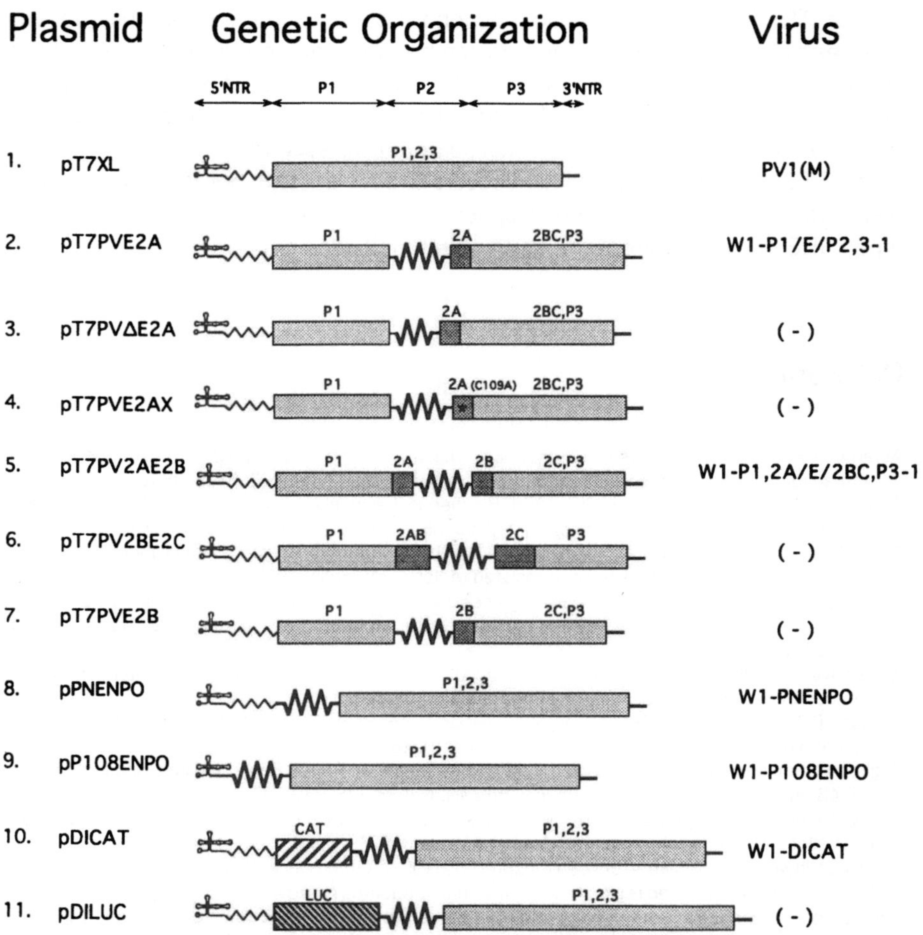

Fig. 8. Genetic organization of dicistronic poliovirus mRNA genomes. The 5' terminal, 108 nt fragment of the PV1 (M) 5' NTR is shown as a *cloverleaf*, the downstream segment of the 5'-NTR (including the IRES) is shown as *thin zig-zag lines*, and the segments of the encephalomyocarditis (EMCV) 5' NTR inserted into the poliovirus genome are shown as *thick zig-zag lines*. The EMCV segments correspond to nt 260-848 (in plasmids 1 and 4), to nt 260-833 (in plasmids 5–11), and to nt 435-833 (in plasmid 3). The *stippled rectangles* represent poliovirus coding regions, and *cross-hatched rectangles* represent chloramphenicol acetyltransferase (CAT) and luciferase (LUC) coding regions, as indicated. Viruses recovered after transfection of mRNA transcripts into HeLa cells are described using standard nomenclature; (–) indicates that a viable virus was not recovered. Adapted from previous reports (ALEXANDER et al. 1994; MOLLLA et al. 1992, 1994; WIMMER et al. 1993)

functionally related and have an intriguing level of structural and sequence similarity. Some conserved RNA sequences and/or structures may correspond to cis-acting elements involved in IRES function, so that there may also be similarities in the mechanism by which the two types or IRES promote initiation. The function of both types of IRES element appears to depend on a cellular 57 kDa polypeptide, which has been identified as the predominantly nuclear hnRNP protein PTB. However, a specific function for p57/PTB in translation has not yet been established. These two groups can be differentiated on the basis of their requirements for trans-acting factors. The EMCV IRES functions efficiently in a broader range of eukaryotic cell types than type 1 IRES elements, probably because the latter require additional factor(s). A second distinction between these IRES element is that initiation occurs directly at the 3' border of type 2 IRES elements, whereas a nonessential spacer of between 30 nt and 154 nt separates type 1 IRES elements from the downstream initiation codon.

References

Abramson RD, Dever TE, Lawson TG, Ray BK, Thach RE, Merrick WC (1987) The ATP-dependent interaction of eukaryotic initiation factors with mRNA. J Biol Chem 262: 3826–3832

Abramson RD, Dever TE, Merrick WC (1988) Biochemical evidence supporting a mechanism for cap-independent and internal initiation of eukaryotic mRNA. J Biol Chem 263:6016–6019.

Adam MA, Ramesh N, Miller AD, Osbourne WRA,(1991) Internal initiation of translation in retroviral vectors carrying picornavirus 5' nontranslated regions. J Virol 65: 4985–4990

Alexander L, Lu HH, Wimmer E (1994) Polioviruses containing picornavirus type 1 and/or type 2 internal ribosomal entry site elements: genetic hybrids and the expression of a foreign gene. Proc Natl Acad Sci USA 91: 1406–1410

Alonso MA, Carrasco L (1981) Reversion by hypotonic medium of the shutoff of protein synthesis induced by encephalomyocarditis virus J Virol 37: 535–540

Anthony DD, Merrick WC (1991) Eukaryotic intiation factor (eIF)-4F Implications for a role in internal initiation of translation. J Biol Chem 266: 10218–10226.

Baglioni C, Simili M, Shafritz DA (1978) Initiation activity of EMCV virus RNA, binding to initiation factor eIF-4B and shut-off of host cell protein synthesis. Nature 275: 240–243

Bandyopadhyay PK, Wang C, Lipton HL (1992) Cap-independent translation by the 5' untranslated region of Theiler's murine encephalomyelitis virus. J Virol 66: 6249–6256

Bandyopadhyay PK, Pritchard A, Jensen K, Lipton HL (1993) A three-nucleotide insertion in the H stem-loop of the 5' untranslated region of Theiler's virus attenuates neurovirulence. J Virol 67: 3691–3695

Beck E, Forss S, Strebel K, Cattaneo R, Feil G (1983) Structure of the FMDV translation initiation site and of the structural proteins. Nucleic Acids Res 11: 7873–7885

Belsham GJ (1992) Dual initiation sites of protein synthesis on foot-and-mouth disease virus RNA are selected following internal entry and scanning of ribosomes in vivo. EMBO J 11: 1105–1110

Belsham GJ, Brangwyn JK (1990) A region of the 5' noncoding region of foot-and-mouth disease virus RNA directs efficient internal initiation of protein synthesis within cells: involvement with the role of L protease in translational control. J Virol 64: 5389–5395

Bennett M, Michaud S, Kingston J, Reed R (1992a) Protein components specifically associated with prespliceosome and spliceosome complexes. Genes Dev 6: 1986–2000

Bennett M, Pinol-Roma S, Staknis D, Dreyfuss G, Reed R (1992b) Differential binding of heterogenous nuclear ribonucleoproteins to mRNA precursors prior to spliceosome assembly in vitro. Mol Cell Biol 12: 3165–3175

Black DN, Stephenson P, Rowlands DJ, Brown F (1979) Sequence and location of the poly (C) tract in aphtho- and cardiovirus RNA. Nucleic Acids Res 6: 2381–2390

Blair GE, Dahl HHM, Truelsen E, Lelong JC (1977) Functional identity of a mouse ascites and a rabbit reticulocyte initiation factor required for natural mRNA translation. Nature 265: 651–653

Borman A, Jackson RJ (1992) Initiation of translation of human rhinovirus RNA: mapping internal ribosomal entry site. Virology 88: 685–696

Borman A, Howell MT, Patton JG, Jackson RJ (1993) The involvement of a spliceosome component in internal initiation of human rhinovirus RNA translation. J Gen Virol 74: 1775–1778

Borovjagin AV, Evstafieva AG, Ugarova TY, Shatsky IN (1990) A factor that specifically binds to the 5'-untranslated region of encephalomyocarditis virus RNA. FEBS Lett 261: 237–240

Borovjagin AV, Ezrokhi MV, Rostapshov VM, Ugarova TY, Bystrova TF, Shatsky IN (1991) RNA-protein

interactions within the internal translation initiation region of encephalomyocarditis virus RNA. Nucleic Acids Res 19: 4999–5005
Borovjagin AV, Pestova TV, Shatsky IN (1994) Pyrimidine tract binding protein strongly stimulates in vitro encephalomyocarditis virus RNA translation at the level of preinitiation complex formation. FEBS Letts 351: 299–302
Bothwell ALM, Ballard DW, Philbrick WM, Lindwall G, Maher SE, Bridgett MM, Jamison SF, Garcia-Blanco MA (1991) Murine polypyrimidine tract binding protein. Purification, cloning, and mapping of the RNA binding domain. J Biol Chem 266: 24657–24663
Brahms J, Maurizot JC, Michelson AM (1967) Conformation and thermodynamic properties of oligo-cytidylic acids. J Mol Biol 25: 465–480
Brown EA, Day SP, Jansen RW, Lemon SM (1991) The 5' nontranslated region of hepatitis A virus RNA: secondary structure and elements required for translation in vitro. J Virol 65: 5828–5838
Brown EA, Zajac AJ, Lemon SM (1994) In vitro characterization of an internal ribosomal entry site (IRES) present within the 5' nontranslated region of hepatitis A virus RNA: comparison with the IRES of encephalomyocarditis virus. J Virol 68: 1066–1074
Brown F, Newman J, Stott J, Porter A, Frisby D, Newton C, Carey N, Fellner P (1974) Poly (C) in animal viral RNAs. Nature 251: 342–344
Brunel F, Alzari PM, Ferrara P, Zakin MM (1991) Cloning and sequencing of a PYBP, a pyrimidine - rich specific single strand DNA binding protein. Nucl Acid Res 19: 5237–5245
Burd, C, Matunis E, Dreyfuss G (1991) The multiple RNA-binding domains of the mRNA poly (A)-binding protein have different RNA-binding activities. Mol Cell Biol 11: 3419–3424
Butterworth BE, Hall L, Stoltzfus CM, Rueckert RR (1971) Virus-specific proteins synthesized in encephalomyocarditis virus-infected cells. Proc Natl Acad Sci USA 68:3083–3087
Canaani D, Revel M, Groner Y (1976) Translational discrimination of 'capped' and 'noncapped' mRNAs: inhibition by a series of chemical analogs of m^{7}GpppX. FEBS Lett 64: 326–331
Chang KH, Brown EA, Lemon SM (1993) Cell type-specific proteins which interact with the 5' nontranslated region hepatitis A virus RNA. J Virol 67: 6716–6725
Chumakov KM, Agol VI (1976) Poly (C) sequence is located near the 5'-end of encephalomyocarditis virus RNA. Biochem Biophys Res Commun 71. 551–557
Chumakov KM, Chichkova NV, Agol VI (1979) 5'-terminal sequence of encephalomyocarditis virus RNA: localization of the poly (C) tract, and role in translation. Dokl Akad Nauk SSSR 246: 994–996
Clarke BE, Sangar DV, Burroughs JN, Newton SE, Carroll AR, Rowlands DJ (1985) Two initiation sites for foot-and mouth disease virus polyprotein in vivo. J Gen virol 66: 2615–2626
Clarke BE, Brown AL, Currey KL, Newton SE, Rowlands DJ, Carroll AR (1987) Potential secondary and tertiary interactions in the genomic RNA of foot and mouth disease virus. Nucleic Acids Res 15: 7067–7079
Coller B-AG, Chapman NM, Beck MA, Pallansch MA, Gauntt CJ, Tracy SM (1990) Echovirus 22 is an atypical enterovirus. J Virol 64: 2692–2701
Dalgarno L, Cox RA, Martin EM (1967) Polyribosomes in normal Krebs 2 ascites tumor cells and in cells infected with encephalomyocarditis virus. Biochim Biophys Acta 138: 316–328
Daniels-McQueen S, Detjen BM, Grifo JA, Merrick WG, Thach RE (1983) Unusual requirements for optimum translation of polio viral RNA in vitro. J Biol Chem 258: 7195–7199
Davies MV, Kaufman RJ (1992) The sequence context of the initiation codon in the encephalomyocarditis virus leader modulates efficiency of internal translation initiation. J Virol 66: 1924–1932
Deng H, Wang C, Ascadi G, Wolff JA (1991) High-efficiency protein synthesis from T7 TNA polymerase transcripts in 3T3 fibroblasts. Gene 109: 193–201
Detjen BM, Jen G, Thach RE (1981) Encephalomyocarditis viral RNA can be translated under conditions of poliovirus-induced translation shutoff in vivo. J Virol 38: 777–781
Dorner AJ, Dorner LF, Larsen GR, Wimmer E, Anderson CW (1982) Identification of the initiation site of poliovirus polyprotein synthesis. J Virol 42: 1017–1028
Dorner AJ, Semler BL, Jackson RJ, Hanecak R, Duprey E, Wimmer E (1984) In vitro translation of poliovirus RNA: utilization of internal initiation sites in reticulocyte lysates. J Virol 50: 507–514
Duke GM, Hoffman MA, Palmenberg AC (1992) Sequence and structural elements that contribute to efficient encephalomyocarditis virus RNA translation. J Virol 66: 1602–1609
Edery I, Lee KAW, Sonenberg N (1984) Functional characterization of eukaryotic mRNA cap binding protein complex: effects on translation of capped and naturally uncapped RNAs. Biochemistry 23: 2456–2462
Elroy-Stein O, Moss B (1990) Cytoplasmic expression system based on constitutive synthesis of bacteriophage T7 RNA polymjerase in mammalian cells. Proc Natl Acad Sci USA 87:6743–6747
Elory-Stein O, Fuerst TR, Moss B (1989) Cap-independent translation of mRNA conferred by encephalo-

myocarditis virus 5' sequence improves the performance of the vaccinia virus/baceteriophage T7 hybrid expression system. Proc Natl Acad Sci USA 86: 6126–6130

Escarmis C, Toja M, Medina M, Domingo E (1992) Modifications of the 5' untranslated region of foot-and-mouth disease virus after prolonged persistence in cell culture. Virus Res 26: 113–125

Etchison D, Smith K (1990) Variations in cap-binding complexes from uninfected and poliovirus-infected HeLA cells. J Biol Chem 265· 358–362

Etchison D, Milburn SC, Edery I, Sonenberg N, Hershey JWB (1982) Inhibition of HeLa cell protein synthesis following poliovirus infection correlates with the proteolysis of a 220,000 dalton polypeptide associated with eukaryotic initiation factor 3 and a cap-binding protein complex. J Biol Chem 257: 14806–14810

Evstafieva AG, Ugarova TY, Chernov BK, Shatsky IN (1991) A complex RNA sequence determines the internal initiation of encephalomyocarditis virus RNA translation. Nucleic Acids Res 19: 665–671

Evstafieva AG, Beletsky AV, Borovjagin AV, Bogdanov AA (1993) Internal ribosomal entry site of encephalomyocarditis virus RNA is unable to direct translation in Saccharmoyces cerevisiae. FEBS Lett 335: 273–276

Forss S, Strebel K, Beck E, Schaller H (1984) Nucleotide sequence and genome organization of foot-and-mouth disease virus. Nucleic Acids Res 12: 6587–6601

Frisby D, Eaton M, Fellner P (1976) Absence of 5'–terminal capping in encephalomyocarditis virus RNA. Nucleic Acids. Res 3: 2771–2787

Garcia-Blanco MA, Jamison SF, Sharp PA (1989) Identification and purification of a 62 000 dalton protein that binds specifically to the polypyrimidine tract of introns. Genes Dev 3: 1874–1886

Ghattas IR, Sanes JS, Majors JE (1991) The encephalomyocarditis virus internal entry site allows efficient coexpression of two genes from a recombinant provirus in cultured cells and in embryos. Mol Cell Biol 11· 5848–5859

Ghetti A, Pinol-Roma S, Michael WM, Morandi C, Dreyfuss G (1992) hnRNP I, the polypyrimidine tract-binding protein: distinct nuclear localization and association with hnRNAs. Nucleic Acids Res. 20: 3671–3678

Gil A, Sharp PA, Jamison SF, Garcia-Blanco MA (1991) Characterization of cDNAs encoding the polypyrimidine tract-binding protein. Genes Dev 5: 1224–1236

Glass MJ, Jia X-Y, Summers DF (1993) Identification of the hepatitis A virus internal ribosomal entry site: in vivo and in vitro analysis of bicistronic RNAs containing the HAV 5' noncoding region. Virology 193: 842–852

Golini F, Thach SS, Birge CH, Safer B, Merrick WC, Thach RE (1976) Competition between cellular and viral mRNAs in vitro is regulated by a messenger discriminatory initiation factor Proc Natl Acad Sci USA 73: 3040–3043

Golini F, Nomoto A, Wimmer E (1978) The genome-linked protein of picornaviruses. IV. Difference in the VPg's of encephalomyocarditis virus and poliovirus as evidence that the genome-linked proteins are virus-coded. Virology 89. 112–118

Hackett PB, Egberts E, Traub P (1978) Selective translation of mengovirus RNA over host mRNA in homologous, fractionated, cell-free translational systems from Ehrlich-ascites-tumor cells. Eur J Biochem 83: 353–361

Hellen CUT, Witherell GW, Schmid M, Shin SH, Pestova TV, Gil A, Wimmer E (1993) A cytoplasmic protein (p57) that is required for translation of picornavirus RNA by internal ribosomal entry is identical to the nuclear pyrimidine-tract-binding protein. Proc Natl Acad Sci USA 90: 7642–7646

Hellen CUT, Pestova TV, Litterst M, Wimmer E (1994) The cellular polypeptide p57 (pyrimidine-tract binding protein) binds to multiple site in the poliovirus 5' nontranslated region. J Virol 68: 941–950

Hershey JWB (1991) Translational control in mammalian cells. Annu Rev Biochem 60: 717–755

Hruby DE, Roberts WK (1978) Encephalomyocarditis virus RNA III. Presence of a genome-associated protein. J Virol 25: 413–415

Hunt SL, Kaminski A, Jackson RJ (1993) The influence of viral coding sequences on the efficiency of internal initiation of translational of cardiovirus RNAs. Virology 197: 801–807

Hyppia T, Horsnell C, Maaronen M, Khan M, Kalkkinen N, Auvinen P, Kinnunen L, Stanway G (1992) A distinct picornavirus group identified by sequence analysis. Proc Natl Acad Sci USA 89: 8847–4451

Jackson RJ (1989) Comparison of encephalomyocarditis virus and poliovirus with respect to translation initiation and processing in vitro. In: Semler BL, Ehrenfeld E (eds) Molecular aspects of picornavirus infection and detection. American Society for Microbiology, Washington

Jackson RJ (1990) Binding of Met–tRNA. In: Trachsel H(ed) Translation in eukaryotes. CRC Press, Boca Raton pp 193–242

Jackson RJ (1991a) Potassium salts influence the fidelity of mRNA translation initiation in rabbit reticulocytes: unique features of encephalomyocarditis virus RNA translation. Biochem Biophys Acta 1088: 345–358
Jackson RJ (1991b) The ATP requirement for initiation of eukaryotic translation varies according to the mRNA species. Eur J Biochem 200: 285–294
Jackson RJ, Howell MT, Kaminski A (1990) The novel mechanism of initiation of picornavirus RNA translation. Trends Biochim Sci 15: 477–483
Jacobson MF, Baltimore D (1968) Polypeptide cleavages in the formation of poliovirus polyproteins. Proc Natl Acad Sci USA 61. 77–84
Jamison SF, Crow A, Garcia-Blanco MA (1992) The spliceosome assembly pathway in mammalian extracts. Mol Cell Biol 12: 4279–4287
Jang SK (1989) Translation of picornaviral mRNAs: initiation of protein synthesis by internal entry of ribosomes into the 5' nontranslated region of picornavirus mRNAs. PhD thesis, State University of New York, Stony Brook
Jang SK, Wimmer E (1990) Cap-independent translation of encephalomyocarditis virus RNA; structural elements of the internal ribosomal entry site and involvement of a cellular 57-kD RNA-binding protein. Genes Dev 4: 1560–1572
Jang SK, Kräusslich H-G, Nicklin MJH, Duke GM, Palmenberg AC, Wimmer E (1988) A segment of the 5' nontranslated region of encephalomyocarditis virus RNA directs internal entry of ribosomes during in vitro translation. J Virol 62: 2636–2643
Jang SK, Davies MV, Kaufman RJ, Wimmer E (1989) Initiation of protein synthesis by internal entry of ribosomes into the 5' nontranslated region of encephalomyocarditis virus RNA in vivo. J Virol 63: 1651–1660
Jang SK, Pestova TV, Hellen CUT, Witherell GW, Wimmer E (1990) Cap-independent translation of picornavirus RNAs: structure and function of the internal ribosomal entry site. Enzyme 44: 292–309
Jen G, Thach RE (1982) Inhibition of host translation in encephalomyocarditis virus-infected L cells: a novel mechanism. J Virol 43: 250–261
Jen G, Birge CH, Thach RE (1978) Comparison of initiation rates of encephalomyocarditis virus and host protein synthesis in infected cells. J Virol 27: 640–647
Jen G, Detjen BM, Thach RE (1980) Shutoff of HeLa cell protein synthesis by encephalomyocarditis virus and poliovirus: a comparative study. J Virol 35. 150–156
Kaminski A, Howell MT, Jackson RJ (1990) Initiation of encephalomyocarditis virus RNA translation: the authentic initiation site is not selected by a scanning mechanism. EMBO J 9: 3753–3759
Kaminski A, Belsham GJ, Jackson RJ (1994) Translation of encephalomyocarditis virus RNA: parameters influencing the selection of the internal initiation site. EMBO J 13: 1673–1681
Kaufman RJ, Davies MV, Walsay LC, Michnick D (1991) Improved vectors for stable expression of foreign genes in mammalian cells by use of the untranslated leader sequence from EMC virus. Nucleic Acids. Res 19: 4485–4490
Kenan DJ, Query CC, Keene J (1991) RNA recognition: towards identifying determinants of specificity. Trends Biochem Sci 16: 214–220
Kim DG, Kang HM, Jang SK, Shin H-S (1992) Construction of a bifunctional mRNA in the mouse by using the internal ribosomal entry site of the encephalomyocarditis virus. Mol Cell Biol 12: 3636–3643
Kong W-P, Roos RP (1991) Alternative translation initiation site in the DA strain of Theiler's murine encephalomyocarditis virus. J Virol 65: 3395–3399
Kozak M (1989) Structural features in eukaryotic mRNAs that modulate the efficiency of translation. J Biol Chem 266: 19867–19870
Kozak M (1991) The scanning model for translation: an update. J Cell Biol 108: 229–241
Kozak M (1992) A consideration of alternative models for the initiation of translation in eukaryotes Crit Rev Biochem Mol Biol 27: 385–402
Kozak M (1994) Features in the 5' non-coding sequences of rabbit α-and β-globin mRNAs that affect translational efficiency. J Mol Biol 235: 95–110
Kräusslich H-G, Nicklin MJH, Toyoda H, Etchison D, Wimmer E (1987) Poliovirus proteinase 2A induces cleavage of eukaryotic initiation factor 4F polypeptide p220. J Virol 61: 2711–2718
Kühn R, Luz N, Beck E (1990) Functional analysis of the internal initiation site of foot-and-mouth disease virus. J Virol 64: 4625–4631
Laskey RA, Gurdon JB, Crawford LV (1972) Translation of encephalomyocarditis viral RNA in oocytes of Xenopus laevis. Proc Natl Acad Sci USA 69: 3665–3669
Law KM, Brown TDK (1990) The complete nucleotide sequence of the GDVII strain of Theiler's murine encephalomyocarditis virus (TMEV). Nucleic Acids Res 18: 6707

Lawrence C, Thach RE (1974) Encephalomyocarditis virus infection of mouse plasmacytoma cells. I. Inhibition of cellular protein synthesis. J Virol 14: 598–610
Le S-Y, Zuker M (1990) Common structures of the 5' non-coding RNA in enteroviruses and rhinoviruses. Thermodynamical stability and statistical significance. J Mol Biol 216: 729–741
Le S-Y, Chen J-H, Sonenberg N, Maizel JV (1993) Conserved tertiary structural elements in the 5' nontranslated region of cardiovirus, aphthovirus and hepatitis A virus RNAs. Nucleic Acids Res 21: 2445–2451
Luz N, Beck E (1990) A cellular 57 kDa protein binds to two regions of the internal translation initiation region of foot-and-mouth disease virus. FEBS Lett 269: 311–314
Luz N, Beck E (1991) Interaction of a cellular 57-kilodalton protein with the internal translation initiation site of foot-and-mouth disease virus. J Virol 65: 6486–6494
Martinez-Salas E, Saiz J-C, Davila M, Belsham GJ, Domingo E (1993) A single nucleotide substitution in the internal ribosome entry site of foot-and-mouth disease virus leads to enhanced cap-independent translation in vivo. J Virol 67: 3748–3755
Mathews MB, Korner A (1970) Mammalian cell-free protein synthesis directed by viral ribonucleic acid. Eur J Biochem 17: 328–338
Meerovitch K, Pelletier J, Sonenberg N (1989) A cellular protein that binds to the 5'-noncoding region of poliovirus RNA: implications for internal translation initiation. Genes Dev 3: 1026–1034
Meerovitch K, Svitkin YV, Lee HS, Lejbkowicz F, Kenan DJ, Chan EKL, Agol VI, Keene JD, Sonenberg N (1993) La autoantigen enhances and corrects aberrant translation of poliovirus RNA in reticulocyte lysate. J Virol 67: 3798–3807
Merrick WC (1992) Mechanism and regulation of eukaryotic protein synthesis. Microbiol Rev 56: 291–315
Michaud S, Reed R, (1991) An ATP-independent complex commits pre-mRNA to the mammalian-spliceosome assembly pathway Genes Dev 5: 2534–2546
Mirochnitchenko O, Inouye S, Inouye M (1994) Production of a single-stranded DNA in mammalian cells by means of a bacterial retron. J Biol Chem 269: 2380–2383
Molla A, Jang SK, Paul AV, Reuer Q, Wimmer E (1993) Cardioviral internal ribosomal entry site is functional in a genetically engineered dicistronic poliovirus. Nature 356: 255–257
Molla A, Paul AV, Schmid M, Jang SK, Wimmer E (1994) Studies on dicistronic polioviruses implicate proteinase $2A^{pro}$ in RNA replication Virology 196: 739–747
Moore M, Query C, Sharp PM (1993) Splicing of precursors to mRNAs by the spliceosome. In: Gestland RF, Atkins JF (eds) The RNA world. Cold Spring Harbor Laboratory Press, Plainview, pp 303–357
Morgan RA, Couture L, Elroy-Stein O, Ragheb J, Moss B, Anderson WF (1992) Retroviral vectors containing putative internal ribosome entry sites: development of a polycistronic gene transfer system and applications to human gene therapy. Nucleic Acids Res 20: 1293–1299
Mosenkis J, Daniels-McQueen S, Janovec S, Duncan R, Hershey JWB, Grifo JA, Merrick WC, Thach RE (1985) Shutoff of host translation by encephalomyocarditis virus infection does not involve cleavage of the eukaryotic initiation factor 4F polypeptide that accompanies poliovirus infection. J Virol 54: 43-645
Mullen MP, Smith CW, Patton JG, Nadal-Ginard B (1991) α-tropomyosin mutually exclusive exon selection: competition between barnchpoint/polypyrimidine tract determines default exon choice. Genes Dev 5: 642–655
Mulligan GJ, Guo W, Wormsley S, Helfman DM (1992) Polypyrimidine tract binding protein interacts with sequences involved in alternative splicing of β-tropomyosin pre-mRNA. J Biol Chem 267: 25480–25487
Newton SE, Carroll AR, Campbell RO, Clarke BE, Rowlands DJ (1985) The sequence of foot-and-month disease virus RNA to the 5'side of the poly(C) tract. Gene 40: 331–336
Nicholson R, Pelletier J, Le S-Y, Sonenberg N (1991) Structural and functional analysis of the ribosome landing pad of poliovirus type 2: in vivo translation studies. J Virol 65: 5886–5894
Nietfeld W, Metzel H, Pieler T (1990) The Xenopus laevis poly(A) binding protein is composed of multiple functionally independent RNA binding domains. EMBO J 9.3699–3705
Norton PA, Hynes RO (1993) Characterization of HeLa nuclear factors which interact with a conditionally processed rat fibronectin pre-mRNA. Biochem Biophys Res Commun 195: 215–221
Ohara Y, Stein S, Lu J, Stillman L, Klaman L, Roos R (1988) Molecular cloning and sequence determination of DA strain of Theiler's murine encephalomyocarditis viruses. Virology 164: 245–255
Oudshoorn P, Thomas A, Scheper G, Voorma HO (1990) An initiation signal in the 5'untranslated leader sequence of encephalomyocarditis virus RNA. Biochim Biophys Acta 1050: 124–128

Palmenberg AC (1989) Sequence alignments of picornaviral capsid proteins. In: Semler BL, Ehrenfeld E (eds) Molecular aspects of picornavirus infection and detection. American Society for Microbiology, Washington

Palmenberg AC, Duke GM (1993) The genomic sequence of mengovirus and its relationship to other cardioviruses. Genebank, no L22089

Palmenberg AC, Kirby EM, Janda MR, Drake NL, Duke GM, Potratz KF, Collett MS (1984) The nucleotide and deduced amino acid sequences of the encephalomyocarditis viral polyprotein coding region. Nucleic Acids Res.12:2969–2985

Parks GD, Duke GM, Palmenberg AC (1986) Encephalomyocarditis 3C protease: efficient cell-free expression from clones which link viral 5' noncoding sequences to the P3 region. J Virol 60: 376–384

Patton JG, Meyer SA, Tempst P, Nadal-Ginard B (1991) Characterization and molecular cloning of polypyrimidine tract-binding protein: a component of a complex necessary for pre-mRNA splicing. Genes Dev.5: 1237–1251

Patton JG, Porro EB, Galceran J, Tempst P, Nadal-Ginard B (1993) Cloning and characterization of PSF, a novel pre-mRNA splicing factor. Genes Dev.7: 393–406

Pause A, Methot N, Svitkin Y, Merrick WC, Sonenberg N (1994) Dominant negative mutants of mammalian translation initiation factor eIF-4A define a critical role for eIF-4F in cap-dependent and cap-independent initiation of translation. EMBO J 13: 1205–1215

Pelham HRB (1978) Translation of encephalomyocarditis virus RNA in vitro yields an active proteolytic processing enzyme. Eur J Biochem 85:457–462

Pelham HRB, Jackson RJ (1976) An efficient mRNA-dependent translation system from reticulocyte lysates Eur J Biochem 67:247–256

Pelletier J, Sonenberg N (1988) Internal initiation of translation of eukaryotic mRNA directed by a sequence derived from poliovirus RNA. Nature 334: 320–325

Pelletier J, Kaplan G, Racaniello VR, Sonenberg N (1988) Translational efficiency of poliovirus mRNA: mapping of inhibitory cis-acting element within the 5' noncoding region. J Virol 62: 2219–2227

Perez-Bercoff R, Gander M (1978) In vitro translation of mengovirus RNA deprived of the terminally-linked (capping?) protein. FEBS Lett 96: 306–312

Perez-Bercoff R, Kaempfer R (1982) Genomic RNA of mengovirus. V. Recognition of common features by ribosomes and eucaryotic initiation factor 2. J Virol 41: 30–41

Pestova TV, Hellen CUT, Wimmer E (1991) Translation of poliovirus RNA: role of an essential cis-acting oligopyrimidine element within the 5'nontranslated region and involvement of a cellular 57-kilodalton protein J Virol 65: 6194–6204

Pestova TV, Hellen CUT, Wimmer E (1994) A conserved AUG triplet in the 5' nontranslated region of poliovirus can function as an initiation codon in vitro and in vivo. Virology 204: 729–737

Pevear DC, Calenoff M, Rozhon E, Lipton HL (1987) Analysis of the complete nucleotide sequence of the picornavirus Theiler's murine encephalomyocarditis virus indicates that it is closely related to cardioviruses. J Virol 61: 1507–1516

Pilipenko EV, Blinov VM, Chernov BK, Dimitrieva TM, Agol VI (1989) Conservation of the secondary structure elements of the 5'-untranslated region of cardio- and aphthovirus RNAs. Nucl Acids Res 17: 5701–5711

Pilipenko EV, Gmyl AP, Maslova SV, Belov GA, Singakov AN, Huang M, Brown TDK, Agol VI (1994) Starting window, a distinct element in the cap-independent internal initiation of translation on picornaviral RNA. J Mol Biol 241: 398–414

Porter A, Carey N, Fellner P (1974) Presence of a larger poly(rC) tract within the RNA of encephalomyocarditis virus. Nature 243: 675–678

Pöyry T, Kinnunen L, Hovi T (1992) Genetic variation in vivo and proposed functional domains of the 5' noncoding region of poliovirus RNA. J Virol 66: 5313–5319

Pritchard AE, Calenoff MA, Simpson S, Jensen K, Lipton HL (1992) A single base deletion in the 5' noncoding region of Theiler's virus attenuates neurovirulence. J Virol 66: 1951–1958

Rivera VM, Walsh JD, Maizel JV (1988) Comparative sequence analysis of the 5' noncoding region of enteroviruses and rhinoviruses. Virology 185: 42–50

Rosen H, Di Segni G, Kaempfer R (1982) Translational control by messenger RNA competition for eukaryotic initiation factor 2. J Biol Chem 257: 946–952

Sangar DV, Black DN, Rowlands DJ, Harris TJR, Brown F (1980) Location of the initiation site for protein synthesis on foot-and-mouth disease virus RNA by in vitro translation of defined fragments of the RNA. J Virol 33: 59–68

Sangar DV, Newton SE, Rowlands DJ, Clarke BE (1987) All foot and mouth disease virus serotypes initiate protein synthesis at two separate AUGs. Nucl Acids Res 15: 3305–3315

Sankar S, Cheah K-C, Porter AG (1989) Antisense oligonucleotide inhibition of encephalomyocarditis virus RNA translation. Eur J Biochem 184: 465–480

Scheper GC, Thomas AAM, Voorma HO (1991) The 5' untranslated region of encephalomyocarditis virus contains a sequence for very efficient binding of eukaryotic initiation factor eIF2/2B. Biochem Biophys Acta 1089: 220–226

Scheper GC, Voorma HO, Thomas AAM (1992) Eukaryotic initiation factors-4E and-4F stimulate 5' cap-dependent as well as internal initiation of protein synthesis. J Biol Chem 267: 7269–7274

Shih DS, Park I-W, Evans CL, Jaynes JM, Palmenberg AC (1987) Effects of cDNA hybridization on translation of encephalomyocarditis virus RNA J Virol 61: 233–2037

Smith AE (1973) The initiation of protein synthesis directed by the RNA from encephalomyocarditis virus. Eur J Biochem 33: 301–313

Smith AE, Marcker KA (1970) Cytoplasmic methionine transfer RNAs from eukaryotes. Nature 226: 607–610

Sonenberg N (1987) Regulation of translation of poliovirus. Adv Virus Res 33: 175–204.

Sonenberg N, Trachsel H, Hecht S, Shatkin AJ (1980) Differential stimulation of capped mRNA translation in vitro by cap binding protein Nature 285· 331–333

Sosnovtsev SV, Onischenko AM, Petrov NA, Kalashnikova TI, Mamaeva NV, Drygin VY, Perevozchikova NA, Vasilenko SK (1993) Sequence of the L fragment of foot-and-mouth disease virus A strain A22/550 Azerbaijan 65 Genbank, no X74812

Stanway G (1990) Structure, function, and evolution of picornaviruses. J Gen Virol 71: 2483–2501

Svitkin YV, Agol VI (1978) Complete translation of encephalomyocarditis virus RNA and faithful cleavage of virus-specific proteins in a cell-free system from Krebs-2 cells. FEBS Lett 87: 7–11

Svitkin YV, Pestova TV, Maslova SV, Agol VI (1988) Point mutations modify the response of poliovirus RNA to a translation initiation factor a comparison of neurovirulent and attenuated strains. Virology 166: 394–404

Svitkin YV, Meerovitch K, Lee HS, Dholakia JN, Kenan DJ, Agol VI, Sonenberg N (1994) Internal translation initiation on poliovirus RNA: further characterization of La function in poliovirus translation in vitro J Virol 68: 1544–1550

Tesar M, Harmon SA, Summers DF, Ehrenfeld E (1991) Hepatitis A virus polyprotein synthesis initiates from two alternative AUG codons. Virology 186: 609–618

Trachsel H (ed) (1991) Translation in eukaryotes Telford, Caldwell

Tsukiyama-Kohara K, Iizuka N, Kohara M, Nomoto A (1992) Internal ribosome entry site within hepatitis C virus RNA J Virol 66: 1476–1483

Ugarova TY (1987) Primary structure of mRNA and the translation strategy of eukaryotes Mol Biol 21: 888–914

Ugarova TY, Siyanova EY, Svitkin YV, Kazachkov YA, Baratova LA, Agol VI (1984) Partial N-terminal amino acid sequences of polypeptides p14 and p12 of encephalomyocarditis virus are identical and correspond to the N-terminus of the polyprotein. FEBS Lett 170: 339–342

Vartapetian AB, Drygin YF, Chumakov KM, Bogdanov AA (1980) The structure of the covalent linkage between proteins and RNA in encephalomyocarditis virus. Nucleic Acid Res 8· 3729–3742

Vartapetian AB, Mankin AS, Skripkin EA, Chumakov KM, Smirnov VD, Bogdanov AA (1983) The primary and secondary structure of the 5'-end region of encephalomyocarditis virus RNA. A novel approach to sequencing long RNA molecules. Gene 26: 189–195

Vennema H, Rijnbrand R, Heijnen L, Horzinek MC, Spaan WJM (1991) Enhancement of the vaccinia virus/phage T7 RNA polymerase expression system using encephalomyocarditis virus 5'-untranslated region sequences. Gene 108: 201–210

Villa-Komaroff L, Guttman N, Baltimore D, Lodish H (1975) Complete translation of poliovirus RNA in a eukaryotic cell-free system. Proc Natl Acad Sci USA 83: 2330–2334

Wigle DT (1973) Purification of a messenger-specific initiation factor from ascites-cell supernatant. Eur J Biochem 35: 11–17

Wigle DT, Smith AE (1973) Specificity in initiation in a fractionated mammalian cell-free system. Nature New Biol 242: 136–140

Wimmer E (1982) Genome-linked proteins of viruses Cell 28: 199–201

Wimmer E, Hellen CUT, Cao X (1993) Genetics of poliovirus. Annu Rev Genet 27: 353–435

Wimmer E, Murdin AD (1991) Hepatitis A virus and the molecular biology of picornaviruses: a case for a new genus of the family picornaviridae. In: Hollinger FB, Lemon SM, Margolis HS (eds) Viral hepatitis and liver disease. Williams and Wilkins, Baltimore

Witherell GW, Gil A, Wimmer E (1993) Interaction of polypyrimidine tract binding protein with the encephalomyocarditis virus mRNA internal ribosomal entry site. Biochemistry 32. 8268–8275

Witherell GW, Wimmer E (1994) Encephalomyocarditis virus internal ribosomal entry site RNA-protein interactions. J Virol 68: 3183–3192
Witherell GW, Wimmer E (1993) Cap-independent translation of picornavirus mRNAs. In: Doefler W, Böm P (eds) Virus strategies. Molecular biology and pathogenesis VCH, Weinheim, pp 237–248
Wood CR, Morris GE, Alderman EM, Fouser L, Kaufman RJ (1991) An internal ribosome binding site can be used to select for homologous recombinants at an immunoglobulin heavy-chain locus. Proc Natl Acad Sci USA 88: 8006–8010
Zhou Y, Giordano TJ, Durbin RK, McAllister WT (1990) Synthesis of functional mRNA in mammalian cells by bacteriophage T3 RNA polymerase. Mol Cell Biol 10: 4529–4537
Zimmerman A, Nelsen-Salz B, Kruppenbacher JP, Eggers HJ (1994) The complete nucleotide sequence and construction of an infectious cDNA clone of a highly virulent encephalomyocarditis virus. Virology 203: 366–372

Anatomy of the Poliovirus Internal Ribosome Entry Site

E. EHRENFELD[1] and B.L. SEMLER[2]

1 Introduction

The positive strand genomic RNAs of picornaviruses present several unique structural features to the metabolic machinery of a eukaryotic cell. Such features include the absence of a 5' terminal m^7G cap group that is usually required for efficient translation, the presence of a small protein (VPg) covalently attached to the 5' end of the viral RNA, an unusually long (600–1200 nts) stretch of 5' non-coding region (5' NCR) sequences upstream of the initiator AUG, the arrangement of these 5' NCR sequences into extensive and complex secondary and tertiary structures, and the presence of multiple AUG codons upstream of the initiator AUG that may place limitations on the ability of cytoplasmic ribosomes to "scan" these sequences prior to selecting the correct initiation codon used for protein synthesis. Compelling evidence for internal entry of ribosomes and RNA-protein interactions at internal sites within the 5' NCR of picornavirus RNAs came from in vitro and cell culture translation studies using dicistronic mRNAs for poliovirus (PELLETIER and SONENBERG 1988, 1989), encephalomyocarditis virus (EMCV; JANG et al. 1988, 1989; MOLLA et al. 1992), foot and mouth disease virus (FMDV;

[1]Department of Molecular Biology and Biochemistry, School of Biological Sciences, University of California Irvine, CA 92717, USA
[2]Department of Microbiology and Molecular Genetics, College of Medicine, University of California, Irvine, CA 92717, USA

BELSHAM and BRANGWYN 1990) and hepatitis A virus (HAV; GLASS et al. 1993; BROWN et al. 1994). The collective results from these experimental approaches demonstrated that downstream cistrons containing a picornavirus 5' NCR could be translated under conditions in which the upstream cistron was prevented from being translated. The conclusion from such studies was that ribosomes bind internally to the 5' NCR of picornavirus RNAs without the usual mode of scanning from the free 5' terminus. This chapter will describe the structure of the poliovirus (PV) 5' NCR and will outline how genetic and biochemical approaches have been used to demonstrate the multifunctional nature of this structure. First, the role of the PV 5' NCR in viral gene expression will be examined. The current model for the structure of the PV 5' NCR will then be presented, followed by a description of insights gained from viable mutants and attenuated viruses containing 5' NCR lesions. Finally, data summarizing the formation of RNA-protein complexes that direct internal binding of ribosomes for initiation of translation will be described.

2 Roles of the 5' NCR in Viral Gene Expression

Chapter 1 in this volume documents the importance of the 5' NCR in the internal ribosome entry and cap-independent translation mechanisms utilized by PV RNAs. The sequences and structures required for these reactions will be the subject of the remainder of this chapter. It is important to note, however, that structural features of the 5' termini of viral RNAs are also required for RNA replication and perhaps genome packaging signals, as well as for translation initiation. The ends of the viral RNAs must provide binding signals for the viral RNA polymerase and any other viral and cellular polypeptides that comprise the replication complex. It is likely that separate domains in the 5' NCR are utilized for the replication and translation functions, although some overlap may exist. ANDINO and coworkers have shown that the 5' terminal 100 nts form a cloverleaf-like structure that serves as the key determinant for polioviral RNA synthesis (ANDINO et al. 1990). This structure binds the viral proteinase-polymerase precursor 3CD and a 36 kDa ribosome-associated cellular protein, both of whose interactions with the viral RNA are required for RNA replication (ANDINO et al. 1993). Mutations that disrupt this complex formation abolished RNA replication but did not affect translation of the RNA, suggesting that the domains regulating the two functions are independent. The construction of viable chimeras in which the internal ribosome entry site (IRES) of one picornavirus is replaced with that of another, as long as the 5' terminal replication signals are retained, will allow finer mapping of the terminal domain. ALEXANDER et al. (1994) generated a chimeric PV genome in which EMCV IRES sequences (nts 260–840) were inserted into a PV cDNA that had been deleted for PV 5' NCR sequences (nts 109–742). Virus particles containing this chimeric RNA were recovered and were shown to have growth characteristics similar (but not identical) to those of wild-type PV. Similar results have been obtained with chimeras constructed between HAV and EMCV sequences (JIA, TESAR, SUMMERS, and EHRENFELD, manuscript in preparation). Thus, both the 5' terminal replication signals and the internal EMCV IRES sequences appear to function relatively independently in the viral RNA, suggesting a functional separation of domains, as indicated in Fig. 1.

Downstream of the 3' boundary of the poliovirus IRES lies a region (nts 640–742) relatively conserved in length (100–104 nts) but hypervariable in nucleotide sequence. No regular motifs in computed secondary structure and no conserved subregions or common features among different serotypes have been identified (TOYODA et al. 1984; POYRY et al. 1992). Results from insertion and deletion mutagenesis showed that this region had little or no effect on virus replication in cultured cells (KUGE and NOMOTO 1987; KUGE et al. 1989; IIZUKA et al. 1989). In fact, it is entirely missing in the human rhinovirus (HRV) genome (CALLAHAN et al. 1985; STANWAY et al. 1984). Its role in the enterovirus life cycle may be related to host cell interactions and/or viral pathogenesis rather than translation initiation or RNA replication.

3 The Structure of the Poliovirus 5' NCR

3.1 Arrangement into Stem-Loop Domains

A useful understanding of the structure of the 5' NCR will require determination of the entire spatial organization of each structural domain that forms a regulatory element. Although this level of analysis of RNA structure is currently beyond available technology, some reasonable progress has been made in inferring secondary structure foldings of the linear RNA sequence. Several different folding procedures (manual and computer-assisted) were applied to predict a secondary

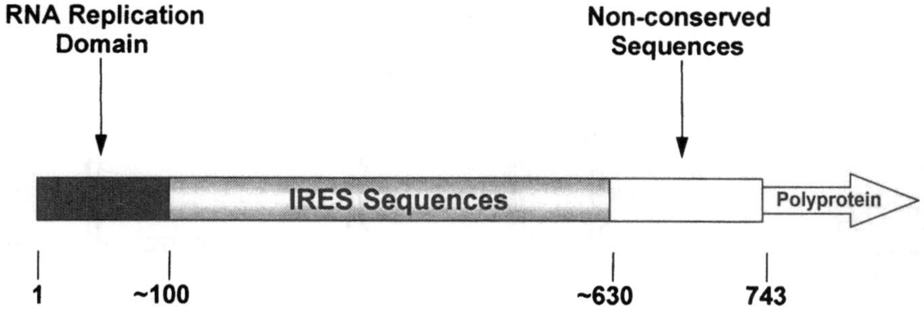

Fig. 1. Functional domains of the 5' NCR of poliovirus RNA. The 5' noncoding region (5' NCR) of poliovirus RNA is depicted by the elongated rectangle preceding the initiation codon (at nt 743) and the NH_2-terminal of the polyprotein (depicted by the horizontal arrow). The RNA replication domain and the translation domain (i.e., IRES sequences) are thought to function, in part, as autonomous elements. It should be emphasized that the junctions between functional domains are not well-defined and the nucleotide numbers shown in the figure (~100 and ~630) are only rough estimates of sequence/function demarcations. Additional information on the different functional domains is provided in the text

structure map of the 5' NCR of PV RNA (RIVERA et al. 1988; PILIPENKO et al. 1989; SKINNER et al. 1989; LE and ZUKER 1990); fortunately all predicted similar overall structures, with some differences in the details of specific regions of individual stem-loop domains. A drawing of a consensus structure is presented in Fig. 2. The nomenclature of the stem-loop domains is taken from HARBER and WIMMER (1993); alternative designations have been utilized by other investigators, as indicated in the figure legend.

The predicted models differ slightly in the folding of the cloverleaf, comprising domain I, of the central region of domain IV, from which several hairpin loops protrude, and of the base or linker regions of domains V and VI. Enzymatic and chemical probings for nucleotides involved in base pairing confirmed many of the structural predictions in domains III (NAJITA and SARNOW 1990), IV and V (SKINNER et al. 1989; PILIPENKO et al. 1989) and VI (PILIPENKO et al. 1989). In addition, an analysis of sequence variation found in independent isolates supported the existence of most of the proposed stem-loop structures by revealing extensive structure-conserving substitutions in the stems (POYRY et al. 1992). A few regions with absolutely conserved sequences, as well as regions with high sequence variability, were observed. These analyses were consistent with the structural predictions and attest to the success in predicting a pattern of RNA structural elements based on the thermodynamics of nucleotide secondary interactions. There have, however, been no direct measurements of the overall RNA structure in the PV 5' NCR; indeed, at present there is little technology developed for physical measurements that lead to the solution of large RNA structures, as are available for comparable analyses of proteins. Although limitations in interpreting highly complex nuclear magnetic resonance spectra limit applications of this

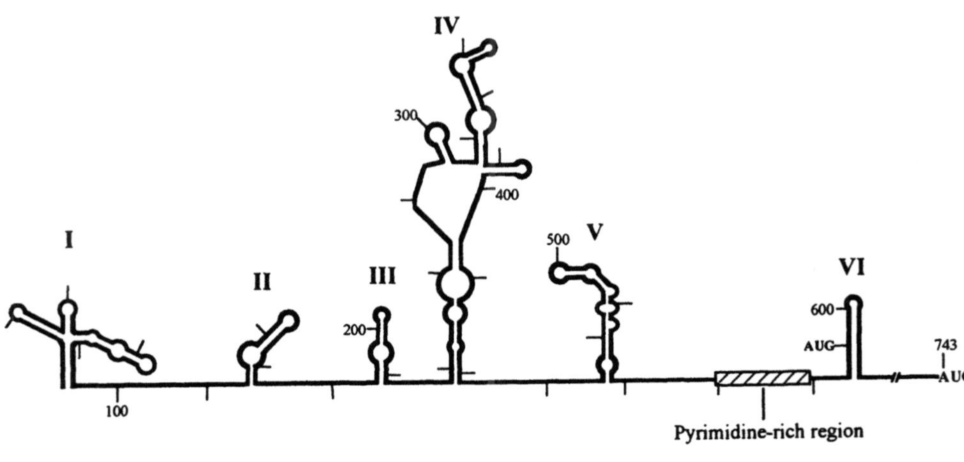

Fig. 2. The predicted stem-loop structures in the poliovirus type 1 5'NCR. The diagram represents a consensus based on previous structural predictions indicated in the text. Domains are numbered according to HARBER and WIMMER (1993). In this representation, domain I is depicted as a cloverleaf composed of the 5' terminal ~100nts. Other authors have designated two separate stem-loop structures near the 5' terminus (I and II or A and B), resulting in a presentation of seven domains, previously called I–VII or A–G

technique to RNA molecules in the range of 50 nts or less, X-ray crystallographic methods may soon increase in application.

The division of the long 5' NCR of PV RNA into domains, as indicated in Fig. 2, has provided an important framework for an experimental approach to dissecting its functions. For example, deletions or other genetic manipulations can be interpreted in terms of their effect on particular stem-loop regions, and subclones of specific cDNAs can be used to generate transcripts representing individual domains which, in turn, provide probes for protein binding studies. At the same time, this division into secondary structure-determined domains poses some constraints on our thinking, since these domains most certainly interact with one another in ways that preclude their functioning as independent units. Thus, a single functional domain or regulatory element may include discontinuous sequences from the linear sequence representation and exclude sequences located in between on the linear map. Nevertheless, there is some evidence that individual domain transcripts can assume structures that are at least similar to functional elements within the intact 5' NCR. For example, transcripts representing only domain VI and parts of the upstream linker region (nts 559–624) can efficiently bind a 52 kDa protein which regulates initiation of translation of RNA sequences from the complete 5' NCR (MEEROVITCH et al. 1989); and transcripts containing only domain IV sequences compete for both translation (BLYN, DILDINE, SEMLER, and EHRENFELD, unpublished observations) of PV RNA and for cross-linking of specific proteins to the full-length 5' NCR (GEBHARD, BLYN, and EHRENFELD, unpublished observations). Similarly, full-length 5' NCR RNAs bind proteins only slightly more efficiently than domain IV fragments, as indicated by competition in mobility shift assays (BLYN, DILDINE, SEMLER, and EHRENFELD, unpublished observations). Despite these examples, smaller RNA fragments have also been shown to bind proteins not bound by the same sequences in the context of the entire 5' NCR, suggesting that a different structural configuration is assumed by the smaller RNA (GEBHARD and EHRENFELD 1992); and some proteins are bound by RNAs containing contiguous sequences from domains V and VI that are not bound by domains V or VI RNAs independently (HALLER and SEMLER 1995).

3.2 Boundaries of the Internal Ribosome Entry Site

Early studies demonstrated that the entire polio 5' NCR was not essential for viability of the virus (KUGE and NOMOTO 1987; KAWAMURA et al. 1989). Deletion of nts 600–726 from the 3' end of the 5' NCR generated virus that grew well, whereas extension of the deletion upstream to nt 564 produced virus with a small plaque phenotype which was assumed to result from impaired translation (KUGE and NOMOTO 1987; PILIPENKO et al. 1992a). The latter mutation involves disruption of domain VI (Fig. 2), and placed the downstream border of the IRES between nts 564 and 600. Other laboratories measured translation efficiencies of reporter RNA constructs in HeLa cell or rabbit reticulocyte lysates which were supplemented with HeLa cell fractions to provide protein factor(s) required for utilization of the IRES. PELLETIER et al. (1988a) observed that removal of nts 632–732 from the 5' NCR had no effect on translation, whereas deletions further upstream (to nt 462) markedly reduced translational activity. Thus, this approach placed the 3' border of the translation element between nts 462 and 632, consistent with the earlier studies of mutant virus growth.

Similar approaches to define the 5' border showed that removal of the first 79 nts had no effect on translation in vitro (PELLETIER et al. 1988a; PELLETIER and SONENBERG 1988), whereas removal of 139 nts slightly reduced translational efficiency. TRONO et al. (1988a) measured translation of a reporter cistron linked to the 5' NCR in transfected cells. Mutations throughout the region between nts 130–600 had detrimental effects on translation, suggesting that most of this region is important either for direct interaction with the translational machinery or to maintain secondary and tertiary structures for this interaction. Thus, the segment of RNA that includes sequences required for cap-independent internal ribosome binding and initiation are confined within the boundaries of nts 130–600, a region which embraces secondary structure domains II–V and part of the sequences in domain VI. Some mutations outside of this region have been reported to influence translation efficiency. For example, linker insertion mutagenesis of the 5'NCR was reported to produce a mutant virus that exhibited a fivefold decrease in viral translation caused by a lesion in the stem-loop formed by nts 10–34 (SIMOES and SARNOW 1991).

3.3 Domains of the Internal Ribosome Entry Site

3.3.1 Domain II

Domain II consists of a simple stem-loop (approx nts 120–165) with one internal bulge. Naturally occurring viral isolates show variations in the size and sequences of the nucleotides comprising the bulge (POYRY et al. 1992). The last base pair of the stem and the loop of the hairpin form a highly conserved motif, C-N-A-N-C-C-A-G, that is repeated in the corresponding position of the stem-loop in domain V and is present in other enteroviruses as well (PILIPENKO et al. 1989). Such repeated and conserved motifs might account for the finding of apparent multiple binding sites for individual proteins (e.g., HELLEN et al. 1994). Point mutations that disrupt the stem or change the loop sequences were debilitating for translation (NICHOLSON et al. 1991). Precise deletion of this stem-loop eliminated expression of a reporter protein used as the second cistron in a bicistronic RNA (PERCY et al. 1992).

3.3.2 Domain III

A stem-loop with an internal bulge constitutes domain III (approx. nts 180–225). Early experiments using polio-coxsackie B3 5' NCR chimeras identified a genetic locus at nt 220 that, upon perturbation, conferred a temperature-sensitive (ts) phenotype to the recombinant virus and had detrimental effects on viral protein synthesis (SEMLER et al. 1986; JOHNSON and SEMLER 1988). Short insertions or deletions at this position in stem-loop III generated viruses with altered phenotypes resulting from impaired translational activity (KUGE and NOMOTO 1987; TRONO et al. 1988b; DILDINE and SEMLER 1989). A revertant virus was recovered, however, with a complete deletion of the stem-loop in domain III (DILDINE and SEMLER 1989). Thus, this structure is not essential for translation. Indeed, precise deletion of this stem-loop in a bicistronic reporter construct had little effect on protein expression in transfected cells (PERCY et al. 1992). Alterations in this domain, when present, might affect the overall tertiary structure of the IRES, however, as suggested by the finding of second site revertants of insertion (at nt 220) mutants within domain V (KUGE and NOMOTO 1987).

3.3.3 Domain IV

This domain (approx. nts 230–445) includes a highly complex structure, with multiple stem-loop segments protruding from a central bulge whose predicted structure differs in detail in the several proposed models. Elimination of the entire domain prevented any protein expression in vivo (PERCY et al. 1992), although translation activity of RNAs with complete or partial deletions in this domain was not abrogated in some in vitro systems (BIENKOWSKA-SZEWCZYK and EHRENFELD 1988; PESTOVA et al. 1989). Internal deletions and small insertions in this region result in viruses with defective translation activity (TRONO et al. 1988 b; KUGE and NOMOTO 1987; NICHOLSON et al. 1991). The locations of these mutations were in several subdomains of this large domain. Interactions between this domain and domain V have been proposed based on the observation of coupled mutations between nt 398 (in domain IV) and nt 481 (in domain V) in PV type 2. These interactions have been postulated to contribute to a potential pseudoknot (PILIPENKO et al. 1992b).

3.3.4 Domain V

This domain (approx. nts 460–540) represents a hairpin structure that includes several internal bulges. Deletion of the entire domain eliminated translation (HALLER and SEMLER 1992; PERCY et al. 1992). Domain V has a key function in translation initiation, since almost any mutation that disrupted base pairing or even partially destroyed the overall structure of this element was lethal to virus replication and to translation in vitro (KUGE and NOMOTO 1987; TRONO et al. 1988b; PELLETIER et al. 1988a; DILDINE and SEMLER 1989; PESTOVA et al. 1989; DILDINE et al. 1991; HALLER and SEMLER 1992). Naturally occurring isolates contain frequent variations in sequence of the terminal stem, always matched by compensating changes that retain the base pairing (POYRY et al. 1992). The loop of the hairpin is the conserved motif also found in the loop of domain II. This domain of the 5' NCR has attracted much attention, since it includes the major determinant of attenuation in the Sabin vaccine strains of PV (EVANS et al. 1985; OMATA et al. 1986; MOSS et al. 1989), and even single point mutations in the stem of this extended hairpin appear to modulate the efficiency of translation (Svitkin et al. 1985, 1990), perhaps in a cell-specific or tissue-specific manner (Agol et al. 1989; La Monica and Racaniello 1989). There is evidence that domain V participates in important tertiary structures (Le et al. 1992) and that its function(s) in protein binding or formation of structural recognition signals may be dependent upon interactions with domain VI (Haller and Semler 1995).

3.3.5 Domain VI

The 3' border of the IRES was identified by deletion and mutational analysis to lie within the stem-loop structure comprising domain VI (approx. nts 580–620). Removal of the 3' half of this hairpin loop had no effect on virus replication (Kuge and Nomoto 1987; Meerovitch et al. 1989) or translation initiation (Meerovitch et al. 1989; Nicholson et al. 1991) nor did other deletions (Percy et al. 1992) or mutations within this region (Haller and Semler 1992; Pilipenko et al. 1992a). By contrast, point mutations in the 5' half of the hairpin loop resulted in marked reductions in translational activity, especially when such lesions were targeted at an AUG triplet (at nt 586), each nucleotide of which is essential. Deletion of the entire structure was lethal (Percy et al. 1992). Thus, the integrity of this structure is not essential for ribosome binding and translation initiation, although portions of the primary sequence and perhaps its ability to participate in higher order interactions with other regions of the 5'NCR are important factors in translational efficiency.

Just upstream of the base of the stem-loop in domain VI, in the linker region between domains V and VI, lies a highly conserved, 21 nt pyrimidine-rich region that is highly sensitive to mutation (Kuge and Nomoto 1987; Iizuka et al. 1989; Nicholson et al. 1991; Meerovitch et al. 1991; Pestova et al. 1991). It has been proposed that this pyrimidine-rich region may hybridize to a complementary segment in 18 S ribosomal RNA (Nicholson et al. 1991; Pilipenko et al. 1992a) (or to a region in 28S ribosomal RNA; Iizuka et al, 1989), although no experimental support for this proposal has been obtained nor has the potential importance of such an interaction been evaluated. The pyrimidine-rich region precedes the essential AUG in the hairpin stem in domain VI by a critical spacing of about 20 nts (Pilipenko et al. 1992a). Site-directed mutagenesis suggests that it is the 5' half of the pyrimidine-rich region (UUUCC at nts 559–563 of PV type 1) that is critical (Meerovitch et al. 1991; Nicholson et al. 1991; Pestova et al. 1991). The conserved spacing between the pyrimidine-rich region and the AUG at the 3' terminus of the IRES suggests that the entire motif constitutes an element that has been postulated to be the site at which ribosome binding occurs (Jackson et al. 1990). Bienkowska-Szewczyk and Ehrenfeld (1988) reported that nts 567–627, which include this conserved element, contained sequences that were essential for efficient translation initiation in a rabbit reticulocyte lysate supplemented with HeLa cell factors. This region was subsequently demonstrated to be the binding site for at least one cellular polypeptide involved in PV translation (Meerovitch et al. 1989, 1993; see below).

4 Insights Gained from Studies on Poliovirus RNAs Containing 5' NCR Lesions

The advent of plasmids containing bacteriophage SP6 or T7 promoters, combined with commercially available phage RNA polymerases, has provided the means for generating highly purified mRNAs containing defined lesions. These RNAs can be used in transfection experiments in cultured cells to determine the effects of mutations on different steps in the virus replication cycle. Using in vitro translation of synthetic mRNAs in rabbit reticulocyte lysate, HeLa cell S-10 extract, or a mixture of both, different investigators have used deletion and mutation analysis to provide evidence for 5' NCR translational requirements in the absence of the potential pleiotropic effects that such lesions may have in vivo. This technology was exploited by NICKLIN et al. (1987) to show that deletion of the first 670 nts of the 5' NCR of PV1 RNA produced a template that was a much better mRNA in rabbit reticulocyte lysate than one containing the intact 5' NCR. A further refinement of the loci that are inhibitory for PV RNA translation in reticulocyte lysates was made by deletion analysis of the type 2 5' NCR fused to a chloramphenicol acetyl transferase (CAT) marker gene (PELLETIER et al. 1988b). These studies narrowed the *cis*-inhibitory locus to between nts 70 and 381 of the viral genome. Of particular interest was the observation that the *cis*-inhibitory domain was not active in HeLa cell extracts (i.e., it did not inhibit translation directed by the PV 5' NCR). This finding was consistent with previous reports that extracts from HeLa cells stimulated translation and suppressed aberrant initiation of protein synthesis when PV RNA was used as mRNA in reticulocyte lysates (BROWN and EHRENFELD 1979; DORNER et al. 1984; PHILLIPS and EMMERT 1986). Taken together, the above studies suggested that HeLa cells may contain factors that interact with the PV 5' NCR to potentiate viral translation, either by relieving the inhibitory effects induced by the sequences between nts 70 and 381 or by specifically interacting with other regions of the 5' NCR to increase binding of proteins required for translation initiation.

Given the ability to recover virus after transfection of cDNA (RACANIELLO and BALTIMORE 1981; SEMLER et al. 1984) or after transfection of RNA made by in vitro synthesis using plasmid DNA templates and bacteriophage RNA polymerases (MIZUTANI and COLONNO 1985; VAN DER WERF et al. 1986), an extensive molecular genetic analysis of picornavirus 5' NCRs has provided evidence for specific sequence requirements in productive viral infections. The effects of mutations of upstream AUG codons in the 5' NCR of PV2 (Lansing) genomic RNA were analyzed by PELLETIER et al. (1988c). In these studies, the seven upstream AUG codons in the PV2 5' NCR were individually converted to UUG codons via site-directed mutagenesis. Only the mutation that affected AUG 7 had a deleterious phenotypic effect on recovered virus. The mutant virus had a small plaque phenotype, produced tenfold lower yields in a one-step growth experiment, and had a slight reduction in viral RNA synthesis (PELLETIER et al. 1988c). Importantly, the mutant displayed a defect in translation that suggested a role for this AUG codon, which is 5' proximal to the initiator AUG used in polyprotein synthesis. As shown in Fig. 2, the AUG at nt 586 is proposed to be part of a base paired stem in the stem-loop VI structure. Following site-directed oligonucleotide mutagenesis, in vitro translation experiments suggested that the sequence of the AUG itself is important within the context of the nts 580–620 stem but not in its contribution to the putative base-paired nature of the stem (MEEROVITCH et al. 1991). Other experiments with infectious cDNA constructs and viable mutant PVs have provided evidence that AUG 586 is recognized within the context of a conserved pyrimidine-rich region that is located just upstream at nts 569–575 (KUGE and NOMOTO 1987; HALLER and SEMLER 1992; PILIPENKO et al. 1992a). JACKSON et al. (1990) have speculated that the AUG at nt 586 (which is conserved in all enteroviruses) may be recognized during initiation of protein synthesis. This initiation event would be important in regulating ribosomal access (perhaps via scanning) to the next AUG downstream (i.e., at nt 743 to initiate polyprotein synthesis) rather than in synthesizing a small peptide, since the reading frame for such a peptide has variable lengths and coding capacities among enteroviruses. In support of the possibility that the 40S ribosomal subunit may scan from the AUG 586 sequence to the start site for polyprotein synthesis, Nomoto and colleagues showed that if an AUG codon was inserted in the highly variable region (located downstream of AUG 586 and proximal to the authentic initiator AUG at nt 743), viruses with small plaque and slow growth phenotypes were recovered (KUGE et al. 1989). These viruses gave rise to large plaque variants, and RNA sequence analysis showed that the genomes of such viruses had either deleted the AUG codon or had mutated one of the three nucleotides in this triplet. Recent work by PESTOVA et al. (1994) showed that AUG 586 could be used as an initiation codon if its surrounding nucleotide sequence was altered by site-directed mutagenesis to conform to the more favorable context found near authentic AUG initiation codons (KOZAK 1989). As described below, the role of AUG 586 in poliovirus translation may involve formation of specific RNA-protein complexes with one or more cellular RNA binding proteins.

Additional lessons learned from viable picornavirus mutants containing lesions in the 5' NCR of genomic RNAs came from the analysis of selected and directed mutations that produce PVs with attenuated neurovirulence phenotypes. It was originally shown by EVANS et al. (1985) that a major determinant of neurovirulence for type 3 PV was found at nt 472 in the 5' NCR. These studies were extended to type 1 (OMATA et al. 1986) and type 2 (MOSS et al. 1989; MACADAM et al. 1991) PVs to show that, while genetic loci in the coding region influenced the attenuation phenotype of the Sabin strains, primary determinants of attenuation/neurovirulence could be mapped to the 5' NCR of viral RNA. Experimental evidence for attenuation/neurovirulence determinants that map to the 5' NCR was generated using viral recombinants (from infectious cDNAs) and pathogenesis studies in monkeys or mice (NOMOTO et al. 1987; LA MONICA et al. 1987; KAWAMURA et al. 1989) or by nucleotide sequence analysis of viruses from stool isolates of vaccinees (EVANS et al. 1985; MINOR and DUNN 1988). In addition, phenotypic analyses of attenuated and neurovirulent polioviruses using cell culture and in vitro translation have provided further evidence for the role of the 5' NCR in the biological properties of these viruses. For example, it has been shown that RNA isolated from Sabin type 3 poliovirions was translated in vitro with reduced efficiency compared to RNAs from neurovirulent strains of type 3 (SVITKIN et al. 1985). The translational deficiency was later conclusively shown to be the result of nt 472 C > U mutation in the 5' NCR of Sabin type 3 genome (SVITKIN et al. 1990). The nt 472 locus in the PV 5' NCR region occurs within the stem-loop V region (refer to Fig. 2). This region of the PV type 1 (Mahoney) genome was a target for linker scanning mutagenesis followed by in vitro translation assays and RNA transfection experiments (HALLER and SEMLER 1992). Results from such experiments suggested that the base-paired regions within stem-loop V are critical to virus translation functions. Interestingly, two pseudorevertant viruses arising from lethal lesions near the attenuation/neurovirulence locus (nt 480) had partially restored base pairing in that region of stem-loop V. These viruses had growth properties and translation efficiencies similar to those of wild-type PV in HeLa cells or in cell-free extracts made from HeLa cells. However, the properties of revertants grown in neuroblastoma cells or translated in a cell-free system containing extracts from neuroblastoma cells were like those of the attenuated Sabin strains, i.e., they grew to lower titers than wild type in neuronal cells and they translated less efficiently than wild type in cell-free extracts from neuronal cells (HALLER and SEMLER, unpublished).

A cell culture model was also used to test the effects of the above 5' NCR mutation at nt 472 on the ability of PVs to replicate in a human neuroblastoma cell line (LA MONICA and RACANIELLO 1989). Recombinant viruses differing only by the nt 472 mutation replicated with equal efficiencies in HeLa cells. However, infection of a neuroblastoma cell line with these two viruses showed that the virus containing the uridine residue at nt 472 grew to a lower titer in these cells, consistent with its attenuation phenotype for mice. The attenuated virus (PRV7.3) also had a reduced translation efficiency during infection of the neural cells, consistent with the above in vitro translation results for Sabin 3 RNAs. Finally, deletion mutagenesis of the 5' NCR of genomic RNA from PV1 showed that deletion of nts 563–727 produced Sabin- and Mahoney-derived viruses that had delayed kinetics of viral translation in cultured cells and had reduced lesion scores in monkey neurovirulence tests (IIZUKA et al. 1989). These data suggest that specific nucleotide sequences in the 5' NCR play an important role in picornavirus pathogenesis, in part, at the level of translation initiation.

A possible role for viral gene product enhancement of IRES-mediated translation initiation has been suggested by transfection experiments with reporter gene constructs linked to the 5' NCR of PV RNA. HAMBIDGE and SARNOW (1992) showed that the presence of a functional 2A proteinase could increase cap-independent translation efficiency of a transfected reporter gene containing the PV 5' NCR. Control experiments suggested that this enhancement was specific for the PV 5' NCR. These results were consistent with those of other investigators who suggested that infection by PV modifies the requirements for specific nucleotide sequences in the viral 5' NCR necessary for internal binding of ribosomes and subsequent initiation of translation (PERCY et al. 1992). It should be noted that these latter studies were carried out in cells infected with recombinant vaccinia viruses, possibly modifying gene expression through nonspecific effects on cap-independent or cap-dependent translation. Interestingly, the same experimental approach (i.e., vaccinia infection/DNA transfection) was employed to provide evidence that the 5' NCR of PV RNA itself could complement translation defects of reporter genes containing defective 5' NCRs of PV (STONE et al. 1993), a result that could also explain the above-mentioned data of PERCY et al. (1992). More recently, the involvement of 2A in cap-independent translation of PV RNA was suggested by the detection of second-site revertants with lesions in 2A that were selected during growth at elevated temperatures of ts viruses containing mismatched base pairs in predicted stem structures in the 5' NCR of type 2 and type 3 RNAs (MACADAM et al. 1994). The above studies have pointed to some fairly unusual mechanisms that PV gene expression may exploit to up-regulate viral-specific, cap-independent translation during an infectious cycle. The precise steps involved , the nature of the cellular effector molecules that may mediate such transactivation, and the cell type dependence (if any) remain to be determined.

5 RNA-Protein Interactions in the 5' NCR of Poliovirus RNA

The presence of internal sequences within picornavirus 5' NCRs that interact directly with cellular components used in protein synthesis led to the search for specific RNA-protein interactions that control translation initiation during picornavirus infections. The initial experimental approaches were based upon electrophoretic mobility shift assays originally developed to analyze RNA splicing mechanisms (KONARSKA and SHARP 1986). In addition, UV cross-linking reactions have been used to further define RNA-protein interactions within the 5' NCR of the poliovirus genome. Such assays were used to show that HeLa cell proteins bind to specific regions of PV RNA (DEL ANGEL et al. 1989; NAJITA and SARNOW 1990; PESTOVA et al. 1991; GEBHARD and EHRENFELD 1992; DILDINE and SEMLER 1992; HALLER and SEMLER 1992; HALLER et al. 1993; MEEROVITCH et al. 1993; HALLER and SEMLER 1995; HELLEN et al. 1994; BLYN, DILDINE, SEMLER, and EHRENFELD, unpublished data).

A number of the RNA-protein interactions described for the PV 5' NCR appear to have a functional relationship to translation initiation. DEL ANGEL et al. (1989) demonstrated that an RNA-protein complex formed with sequences proximal to and including stem-loop II (nts 97–182) was comprised of a number of distinct cellular proteins that included eukaryotic initiation factor 2a (eIF-2a). HELLEN et al. (1994) recently presented evidence that a 57 kDa protein from Hela cells can be UV cross-linked to a PV transcript containing the nts 70–288 RNA sequence (which includes part of stem-loop I, all of stem-loops II and III, and part of stem-loop IV). The 57 kDa protein is identical to the splicing factor, polypyrimidine tract binding protein (PTB) (HELLEN et al. 1993) and, as will be described below, binds to other sequence elements within the PV 5' NCR (HELLEN et al. 1994). This same polypeptide has been identified in reticulocyte lysates, and it binds specifically to the nts 403–447 stem-loop structure of EMCV RNA (JANG and WIMMER 1990). The 57 kDA polypeptide (PTB) may be identical to a 58 kDa protein isolated from Krebs-2 ascites cells that binds to the nts 315–485 region of EMCV RNA (BOROVJAGIN et al. 1990) and to a 57 kDa protein that binds to the analogous region of the 5'NCR of FMDV RNA (LUZ and BECK 1990).

For stem-loop III (nts 186–220), RNA-protein complexes were shown by electrophoretic mobility shift assays and by UV cross-linking to involve a 50 kDa cellular protein that may be membrane-associated (NAJITA and SARNOW 1990; DILDINE and SEMLER 1992). The binding site for the 50 kDa protein was mapped to the single-stranded loop region at the top of stem-loop III (NAJITA and SARNOW 1990). Additional studies showed that the RNA-protein interaction directed by stem-loop III required the presence of a correctly base-paired stem structure but not a specific nucleotide sequence within the stem itself (DILDINE and SEMLER 1992). It is of interest that a previous report had described a deletion of the stem-loop III hairpin structure (refer to Fig. 2) in the 5' NCR of PV1 RNA that relieved a ts phenotype of a mutant with a four nt deletion at nts 221–224 (DILDINE and SEMLER 1989). In addition, this structure is deleted in the nucleotide sequence of bovine enterovirus (EARLE et al. 1988). Thus, the stem-loop III RNA-protein interaction does not appear to be absolutely required for virus growth in cell culture. However, the revertant PV that had deleted the stem-loop III sequences did not have wild-type growth properties. In addition, mRNAs from reconstructed templates containing the stem-loop III deletion reproducibly directed in vitro translation at levels 50%–90% of those produced by RNAs containing wild-type PV sequences. Perhaps the role of RNA-protein interactions encoded by stem-loop III is indirect, either in the formation of higher order structures needed for maximal translation efficiency or in facilitating functions that require association with cellular membranes e.g., viral RNA replication (see SEMLER et al. 1988 for review).

Different complexes between cellular proteins and regions of the 5' NCR thought to be crucial for translation initiation (i.e. stem-loops IV and V) have been detected in mobility shift and UV cross-linking assays. GEBHARD and EHRENFELD (1992) used UV cross-linking of sequentially truncated PV 5'NCR transcripts to identify two proteins (M_r 38 kDa and 48 kDa) that bound to stem-loop IV. Deletion analysis of stem-loop IV sequences revealed that much of the extensive predicted RNA secondary structure in this region of PV RNA is required for formation of specific RNA-protein complexes (DILDINE and SEMLER 1992). For stem-loop V, there is evidence that the isolated, intact secondary structure may not bind cellular proteins in a functionally significant manner. No specific RNA-protein complexes could be reproducibly detected by mobility shift assays when an RNA corresponding to a complete (predicted) stem-loop V sequence (nt 448–556) was incubated with crude extracts from HeLa or cultured neuroblastoma cells (HALLER and SEMLER 1995). However, when stem-loop V sequences were incubated with purified PTB (p57) or cytoplasmic extracts from human cells, a complex with a 57 kDa protein could be detected by UV cross-linking (HELLEN et al. 1994). It is possible that interaction between adjacent stem-loop structures within the PV 5' NCR creates binding sites for cellular proteins not present in the separate forms of such structures. In support of this possibility, a recent report describes the binding of a 36 kDa protein to an RNA transcript containing stem-loops V and VI (HALLER and SEMLER 1995). This protein was not bound to template RNAs containing either stem-loop V alone or stem-loop VI alone. Although a functional role for the 36 kDa protein was not ascertained in the above-mentioned study, it is tempting to speculate that the higher order structure of the PV 5' NCR, resulting from different domains of the RNA interacting with each other, produces novel protein binding sites that may facilitate internal ribosome entry.

An RNA-protein complex specific for the nts 559–624 region of the PV 5' NCR (stem-loop VI) was shown to contain a cellular polypeptide (p52) that was not a known translation factor (MEEROVITCH et al. 1989). This protein was abundant in extracts from HeLa cells but was present in limiting quantities in reticulocyte lysates or wheat germ extracts. A separate study identified a~54 kDa protein with properties very similar to those of p52, suggesting that the two are, in fact, the same polypeptide (GEBHARD and EHRENFELD 1992). These data are consistent with the above-mentioned findings that PV RNA is translated efficiently in extracts from HeLa cells but not in reticulocyte lysates. More recent studies have shown that p52 is identical to the La autoantigen, a protein that is involved in processing RNA polymerase III transcripts (MEEROVITCH et al. 1993). Addition of purified p52/La to in vitro translation reactions using the rabbit reticulocyte lysate increases the efficiency of translation initiation at the authentic AUG start codon of PV mRNA and "corrects" the aberrant initiation products normally observed in this translation system.

The significance of the RNA-protein interactions directed by stem-loop VI warrants further consideration in light of other biochemical and functional data. It had been previously shown that a PV 5' NCR truncated prior to the stem-loop VI sequences (i.e., deletion of nt 1–566) could provide the required sequence determinants for in vitro translation of poliovirus RNA in a rabbit reticulocyte lysate supplemented with an extract from uninfected HeLa cells (BIENKOWSKA-SZEWCZYK and EHRENFELD 1988). The 5' sequences of the truncated transcript still contained part of the conserved pyrimidine-rich region, whose precise spacing (~20–30 nts) from a downstream AUG codon was shown to be crucial for translation initiation and PV infectivity (HALLER and SEMLER 1992; PILIPENKO et al. 1992a; JANG et al. 1990; PESTOVA et al. 1991). Deletion mutations (either directed or selected) which remove all or part of the stem-loop VI sequences are only functional in translation if the proper spacing between the pyrimidine-rich region and a downstream AUG has been restored. Thus, the formation of translation initiation complexes with this region of the PV 5' NCR may involve the recognition by La (p52) and other cellular proteins (DEL ANGEL et al. 1989; HALLER and SEMLER 1992; GEBHARD and EHRENFELD 1992; MEEROVITCH et al. 1989, 1993; HALLER and SEMLER 1995) of two sequence elements (the pyrimidine-rich region and a downstream AUG) separated by about 20–30 nucleotides. Such a model predicts that a single RNA binding protein will have two different domains that contact the RNA or that two proteins with individual RNA binding domains interact with each other via a protein–protein interaction that may be facilitated by individual interactions with the specific RNA sequences. Ultimately, these complexes must be recognized by the translation apparatus of the cell to generate a functional initiation complex that has now bypassed the need for a 5' terminal cap structure and the limitations conferred by a translation mechanism that relies on scanning of a ribosome through all 5' proximal sequences that are upstream of an authentic AUG codon.

6 Concluding Remarks

The work summarized in this chapter represents only a preliminary approach to understanding the interaction between the PV 5' NCR and cellular proteins that mediate translation initiation from the PV IRES. Although members of all genera in the Picornaviridae family direct synthesis of their proteins from an IRES, these elements are very different in sequence and predicted structure among the different groups. For example, the entero- and rhinoviruses share considerable conservation of IRES structure, as do the cardio- and aphthoviruses; however, there is almost no similarity in the pattern of motifs between these two groups. It is not yet clear whether the hepatoviruses contain an independent type of IRES structure, or one that is related to the cardio/aphtho virus element. One conserved feature present in all of the picornavirus IRES elements is the pyrimidine-rich tract about 25 nts upstream of an AUG. This feature may represent the actual site of ribosome entry and thus have the same function in all picornaviral translation initiation reactions. Other portions of the IRES structure, however, may reflect specific interactions with different cellular factors. A full understanding of this complex reaction will ultimately require identification and characterization of all of the *trans*-acting factors involved, as well as a complete picture of the *cis*-acting elements and their spatial organization within the RNA. It may be anticipated that the spectrum of mRNA-specific, cell-specific, and general translation factors utilized by viral RNAs will match the complexity of the transcription factor pattern required to regulate gene expression from viral DNAs.

Acknowledgments. Work described from the authors' laboratories was supported by US Public Health Service grants (AI 12387 and AI 26765) from the National Institutes of Health.

References

Agol VI, Drozdov SG, Ivannikova TA, Kolesnikova MS, Korolev MB, Tolskaya EA (1989) Restricted growth of attenuated poliovirus strains in cultured cells of a human neuroblastoma. J Virol 63: 4034–4038

Alexander L, Lu HH, Wimmer E (1994) Polioviruses containing picornavirus type 1 and/or type 2 internal ribosomal entry site elements: genetic hybrids and the expression of a foreign gene. Proc Natl Acad Sci 91: 1406–1410

Andino R, Rieckhof GE, Baltimore D (1990) A functional ribonucleoprotein complex forms around the 5' end of poliovirus RNA. Cell 63: 369–380
Andino R, Rieckhof GE, Achacoso PL, Baltimore D (1993) Poliovirus RNA synthesis utilizes an RNP complex formed around the 5'-end of viral RNA EMBO J 12: 3587–3598
Belsham GJ, Brangwyn JK (1990) A region of the 5' noncoding region of foot-and-mouth disease virus RNA directs efficient internal initiation of protein synthesis within cells: Involvement with the role of L protease in translational control. J Virol 64: 5389–5395
Bienkowska-Szewczyk K, Ehrenfeld E (1988) An internal 5'-noncoding region required for translation of poliovirus RNA in vitro J Virol 62: 3068–3072
Borovjagin AV, Evstafieva AG, Ugarova TY, Shatsky IN (1990) A factor that specifically binds to the 5'-untranslated region of encephalomyocarditis virus RNA. FEBS Lett 261: 237–240
Brown BA, Ehrenfeld E (1979) Translation of poliovirus RNA in vitro: changes in cleavage pattern and initiation sites by ribosomal salt wash Virology 97: 396–405
Brown EA, Zajac AJ, Lemon SM (1994) In vitro characterization of an internal ribosomal entry site (IRES) present within the 5' nontranslated region of hepatitis A virus RNA: comparison with the IRES of encephalomyocarditis virus. J Virol 68: 1066–1074
Callahan PL, Mizutani S, Colonno RJ (1985) Molecular cloning and complete sequence determination of the RNA genome of human rhinovirus type 14. Proc Natl Acad Sci USA 82: 732–736
del Angel RM, Papavassiliou AG, Fernandez-Thomas C, Silverstein SJ, Racaniello VR (1989) Cell proteins bind to multiple sites within the 5'-untranslated region of poliovirus RNA. Proc Natl Acad Sci USA 86: 8299–8303
Dildine SL, Semler BL (1989) The deletion of 41 proximal nucleotides reverts a poliovirus mutant containing a temperature-sensitive lesion in the 5' noncoding region of genomic RNA. J Virol 63: 847–862
Dildine SL, Semler BL (1992) Conservation of RNA-protein interactions among picornaviruses. J Virol 66: 4364–4376
Dildine SL, Stark KR, Haller AA, Semler BL (1991) Poliovirus translation initiation. differential effects of directed and selected mutations in the 5' noncoding region of viral mRNAs. Virology 182: 742–752
Dorner AJ, Semler BL, Jackson RJ, Hanecak R, Duprey E, Wimmer E (1984) In vitro translation of poliovirus RNA: utilization of internal initiation sites in reticulocyte lysate. J Virol 50: 507–514
Earle JAP, Skuce RA, Fleming CS, Hoey EM, Martin SJ (1988) The complete nucleotide sequence of a bovine enterovirus. J Gen Virol 69 253–263
Evans DMA, Dunn G, Minor PD, Schild GC, Cann AJ, Stanway G, Almond JW, Currey K, Maizel JV (1985) Increased neurovirulence associated with a single nucleotide change in a noncoding region of the Sabin type 3 poliovaccine genome. Nature 314. 548–550
Gebhard JR, Ehrenfeld E (1992) Specific interactions of HeLa cell proteins with proposed translation domains of the poliovirus 5' noncoding region. J Virol 66. 3101–3109
Glass MJ, Jia XY, Summers DF (1993) Identification of the hepatitis A virus internal ribosome entry site: in vivo and in vitro analysis of bicistronic RNAs containing the HAV 5' noncoding region. Virology 193: 842–852
Haller AA, Semler BL (1992) Linker scanning mutagenesis of the internal ribosome entry site of poliovirus RNA. J Virol 66: 5075–5086
Haller AA, Semler BL (1995) Stem-loop structure synergy in binding cellular proteins to the 5' noncoding region of poliovirus RNA. Virology 206: 923–934
Haller AA, Nguyen JHC, Semler BL (1993) Minimum internal ribosome entry site required for poliovirus infectivity. J Virol 67. 7461–7471
Hambidge SJ, Sarnow P (1992) Translational enhancement of the poliovirus 5' noncoding region mediated by virus-encoded polypeptide 2A. Proc Natl Acad Sci USA 89: 10272–10276
Harber J, Wimmer E (1993) Aspects of the molecular biology of poliovirus replication. In: Carrasco L, Sonenberg N, Wimmer E (eds) Regulation of gene expression in animal viruses. Plenum, New York, pp 189–224
Hellen CUT, Witherell GW, Schmid M, Shin SH, Pestova TV, Gil A, Wimmer E (1993) A cytoplasmic 57 kDa protein (p57) that is required for translation of picornavirus RNA by internal ribosomal entry is identical to the nuclear polypyrimidine tract-binding protein. Proc Natl Acad Sci USA 90: 7642–7646
Hellen CUT, Pestova TV, Litterst M, Wimmer E (1994) The cellular polypeptide p57 (pyrimidine tract-binding protein) binds to multiple sites in the poliovirus 5' nontranslated region. J Virol 68: 941–950
Iizuka N, Kohara A, Hagino-Yamagishi K, Abe S, Komatsu T, Tago K, Arita M, Nomoto A (1989) Construction of less neurovirulent polioviruses by introducing deletions into the 5' noncoding sequence of the genome. J Virol 63: 5354–5363

Jackson RJ, Howell MT, Kaminski A (1990) The novel mechanism of initiation of picornavirus RNA translation. TIBS 15: 477–483

Jang SK, Wimmer E (1990) Cap-independent translation of encephalomyocarditis RNA: structural elements of the internal ribosomal entry site and involvement of a cellular 57 kD RNA-binding protein. Genes Dev 4: 1560–1572

Jang SK, Kräusslich HG, Nicklin MJH, Duke GM, Palmenberg AC, Wimmer E (1988) A segment of the 5' nontranslated region of encephalomyocarditis virus RNA directs internal entry of ribosomes during in vitro translation. J Virol 62: 2636–2643

Jang SK, Davies MV, Kaufman RJ, Wimmer E (1989) Initiation of protein synthesis by internal entry or ribosomes into the 5' nontranslated region of encephalomyocarditis virus RNA in vivo. J Virol 63: 1651–1660

Jang SK, Pestova TV, Hellen CUT, Witherell GW, Wimmer E (1990) Cap-independent translation of picornavirus RNAs: structure and function of the internal ribosomal entry site. Enzyme 44: 292–309

Johnson VH, Semler BL (1988) Defined recombinants of poliovirus and coxsackievirus: sequence specific deletions and functional substitutions in the 5'-noncoding regions of viral RNAs. Virology 162: 47–57

Kawamura N, Kohara M, Abe S, Komatsu T, Tago K, Arita M, Nomoto A (1989) Determinants in the 5' noncoding region of poliovirus Sabin 1 RNA that influence the attenuation phenotype. J Virol 63: 1302–1309

Konarska MM, Sharp PA (1986) Electrophoretic separation of complexes involved in the splicing of precursors to mRNAs. Cell 46: 845–855

Kozak M (1989) The scanning model for translation: an update. J Cell Biol 108: 229–241

Kuge S, Nomoto A (1987) Construction of viable deletion and insertion mutants of the Sabin strain type 1 poliovirus: function of the 5' noncoding sequence in viral replication. J Virol 61: 1478–1487

Kuge S, Kawamura N, Nomoto A (1989) Genetic variation occurring on the genome of an in vitro insertion mutant of poliovirus type 1. J Virol 63: 1069–1075

La Monica N, Almond JW, Racaniello VR (1987) A mouse model for poliovirus neurovirulence identifies mutations that attenuate the virus for humans. J Virol 61: 2917–2920

La Monica N, Racaniello VR (1989) Differences in replication of attenuated and neurovirulent poliovirus in human neuroblastoma cell line SH-SY5Y. J Virol 63: 2357–2360

Le SY, Zuker M (1990) Common structures of the 5'-noncoding RNA in enteroviruses and rhinoviruses. J Mol Biol 216: 726–741

Le SY, Chen JH, Sonenberg N, Maizel JV (1992) Conserved tertiary structure elements in the 5' untranslated region of human enteroviruses and rhinoviruses. Virology 191: 858–866

Luz N, Beck E (1990) A cellular 57 kDa protein binds to two regions of the internal translation initiation site of foot-and-mouth disease virus. FEBS Lett 269: 311–314

Macadam AJ, Pollard SR, Ferguson G, Dunn G, Skuce R, Almond JW, Minor PD (1991) The 5' noncoding region of the type 2 poliovirus vaccine strain contains determinants of attenuation and temperature sensitivity. Virology 181: 451–458

Macadam AJ, Ferguson G, Fleming T, Stone DM, Almond JW, Minor PD (1994) Role for poliovirus protease 2A in cap independent translation. EMBO J 13: 924–927

Meerovitch K, Pelletier J, Sonenberg N (1989) A cellular protein that binds to the 5' noncoding region of poliovirus RNA: implications for internal translation initiation. Genes Dev 3: 1026–1034

Meerovitch K, Nicholson R, Sonenberg N (1991) In vitro mutational analysis of cis-acting RNA translational elements within the poliovirus type 2 5' untranslated region. J Virol 65: 5895–5901

Meerovitch K, Lee HS, Svitkin Y, Kenan DJ, Chan EKL, Agol, VI, Keene JD, Sonenberg N (1993) La autoantigen enhances and corrects translation of poliovirus RNA in reticulocyte lysate. J Virol 67: 3798–3807

Minor PD, Dunn G (1988) The effect of sequences in the 5' non-coding region on the replication of polioviruses in the human gut. J Gen Virol 69: 1091–1096

Mizutani S, Colonno RJ (1985) In vitro synthesis of an infectious RNA from cDNA clones of human rhinovirus type 14. J Virol 56: 628–632

Molla A, Jang SK, Paul AV, Reuer Q, Wimmer E (1992) Cardioviral internal ribosomal entry site is functional in a genetically engineered dicistronic poliovirus. Nature 356: 255–257

Moss EG, O'Neill RE, Racaniello VR (1989) Mapping of attenuating sequences of an avirulent poliovirus type 2 strain. J Virol 63: 1884–1890

Najita L, Sarnow P (1990) Oxidation-reduction sensitive interaction of a cellular 50-kDa protein with an RNA hairpin in the 5'-noncoding region of the poliovirus genome. Proc Natl Acad Sci USA 87: 5846–5850

Nicholson R, Pelletier J, Le SY, Sonenberg N (1991) Structural and functional analysis of the ribosome landing pad of poliovirus type 2: in vivo translation studies. J Virol 65: 5886–5894
Nicklin MJH, Kräusslich HG, Toyoda H, Dunn JJ, Wimmer E (1987) Poliovirus polypeptide precursors: expression in vitro and processing by exogenous 3C and 2A proteinases. Proc Natl Acad Sci USA 84: 4002–4006
Nomoto A, Kohara M, Kuge S, Kawamura N, Arita M, Komatsu T, Abe S, Semler BL, Wimmer E, Itoh H (1987) Study on virulence of poliovirus type 1 using in vitro modified viruses. In: Brinton MA, Rueckert RR (eds) Positive strand RNA viruses. Liss, New York, pp 437–452
Omata T, Kohara M, Kuge S, Komatsu T, Abe S, Semler BL, Kameda A, Itoh H, Arita M, Wimmer E, Nomoto A (1986) Genetic analysis of the attenuation phenotype of poliovirus type 1. J Virol 58: 348–358
Pelletier J, Sonenberg N (1988) Internal initiation of translation of eukaryotic mRNA directed by a sequence derived from poliovirus RNA. Nature 334: 320–325
Pelletier J, Sonenberg N (1989) Internal binding of eukaryotic ribosomes on poliovirus RNA: translation in Hela cell extracts. J Virol 63: 441–444
Pelletier J, Kaplan G, Racaniello VR, Sonenberg N (1988a) Cap-independent translation of poliovirus mRNA is conferred by sequence elements within the 5' noncoding region. Mol Cell Biol 8: 1103–1112
Pelletier J, Kaplan G, Racaniello VR, Sonenberg N (1988b) Translational efficiency of poliovirus mRNA: mapping inhibitory cis–acting elements within the 5' noncoding region. J Virol 62: 2219–2227
Pelletier J, Flynn ME, Kaplan G, Racaniello V, Sonenberg N (1988c) Mutational analysis of upstream AUG codons of poliovirus RNA. J Virol 62: 4486–4492
Percy N, Belsham GJ, Brangwyn JK, Sullivan M, Stone DM, Almond JW (1992) Intracellular modifications induced by poliovirus reduce the requirement for structural motifs in the 5' noncoding region of the genome involved in internal initiation of protein synthesis. J Virol 66: 1695–1701
Pestova TV, Maslova SV, Potapov VK, Agol VI (1989) Distinct modes of poliovirus polyprotein initiation in vitro. Virus Res 14: 107–118
Pestova TV, Hellen CUT, Wimmer E (1991) Translation of poliovirus RNA: the essential roles of a cis-acting oligopyrimidine element within the 5'-nontranslated region and a trans-acting 57 kDa protein. J Virol 65: 6194–6204
Pestova TV, Hellen CUT, Wimmer E (1994) A conserved AUG triplet in the 5' nontranslated region of poliovirus can function as an initiation codon in vitro and in vivo Virology 204: 729–737
Phillips BA, Emmert A (1986) Modulation of expression of poliovirus proteins in reticulocyte lysates. Virology 148: 255–267
Pilipenko EV, Blinov VM, Romanova LI, Sinyakov AN, Maslova SV, Agol VI (1989) Conserved structural domains in the 5'-untranslated region of picornaviral genomes: an analysis of the segment controlling translation and neurovirulence. Virology 168: 201–209
Pilipenko EV, Gmyl AP, Maslova SV, Svitkin YV, Sinyakov AN, Agol VI (1992a) Prokaryotic like cis elements in the cap-independent internal initiation of translation on picornavirus RNA. Cell 68: 119–131
Pilipenko EV, Maslova SV, Sinyakov AN, Agol VI (1992b) Towards identification of cis-acting elements involved in the replication of enterovirus and rhinovirus RNAs. a proposal for the existence of tRNA like terminal structures. Nucleic Acids Res 20: 1739–1745
Poyry T, Kinnunen L, Hovi T (1992) Genetic variation in vivo and proposed functional domains of the 5' noncoding region of poliovirus RNA. J Virol 66: 5313–5319
Racaniello VR, Baltimore D (1981) Cloned poliovirus complementary DNA is infectious in mammalian cells. Science 214: 916–919
Rivera VM, Welsh JD, Maizel JV (1988) Comparative sequence analysis of the 5' noncoding region of the enteroviruses and rhinoviruses. Virology 165: 42–50
Semler BL, Dorner AJ, Wimmer E (1984) Production of infectious poliovirus from cloned cDNA is dramatically increased by SV 40 transcription and replication signals. Nucleic Acids Res 12: 5123–5141
Semler BL, Johnson VH, Tracy S (1986) A chimeric plasmid from cDNA clones of poliovirus and coxsackievirus produces a recombinant virus that is temperature-sensitive. Proc Natl Acad Sci USA 83: 1777–1781
Semler BL, Kuhn RJ, Wimmer E (1988) Replication of the poliovirus genome. In: Domingo E, Holland J, Ahlquist P (eds) RNA genetics, vol I. CRC Press, Boca Raton, pp 23–48
Simoes EA, Sarnow P (1991) An RNA hairpin at the extreme 5' end of the poliovirus RNA genome modulates viral translation in human cells. J Virol 65 913–921

Skinner MA, Racaniello VR, Dunn G, Cooper J, Minor PD, Almond JW (1989) New model for the secondary structure of the 5' non-coding RNA of poliovirus is supported by biochemical and genetic data that also shows that RNA secondary structure is important in neurovirulence. J Mol Biol 207: 379–392

Stanway G, Hughes P, Mountford RC, Minor PD, Almond JW (1984) The complete nucleotide sequence of a common cold virus: human rhinovirus 14. Nucleic Acids Res 12: 7859–7875

Stone DM, Almond JW, Brangwyn JK, Belsham GJ (1993) Trans complementation of cap-independent translation directed by poliovirus 5' noncoding region deletion mutants: evidence for RNA–RNA interactions. J Virol 67: 6215–6223

Svitkin YV, Maslova SV, Agol VI (1985) The genomes of attenuated and virulent poliovirus strains differ in their in vitro translation efficiencies. Virology 147. 243–252

Svitkin YV, Cammack N, Minor PD, Almond JW (1990) Translation deficiency of the Sabin type 3 poliovirus genome: association with an attenuating mutation C_{472}→U. Virology 175: 103–109

Toyoda H, Kohara M, Kataoka Y, Suganuma T, Omato T, Imura N, Nomoto A (1984) Complete nucleotide sequences of all three poliovirus serotype genomes. J Mol Biol 174: 561–585

Trono D, Pelletier J, Sonenberg N, Baltimore D (1988a) Translation in mammalian cells of a gene linked to the poliovirus 5' noncoding region. Science 241: 445–448

Trono D, Andino R, Baltimore D (1988b) An RNA sequence of hundreds of nucleotides at the 5' end of poliovirus RNA is involved in allowing viral protein synthesis. J Virol 62: 2291–2299

van der Werf S, Bradley J, Wimmer E, Studier FW, Dunn JJ (1986) Synthesis of infectious poliovirus RNA by purified T7 RNA polymerase. Proc Natl Acad Sci USA 83: 2330–2334.

The Role of the La Autoantigen in Internal Initiation

G.J. Belsham[1,3], N. Sonenberg[1,2], and Y.V. Svitkin[1,4]

1 Introduction

The picornavirus family is widespread in nature and these viruses replicate in a wide variety of hosts and tissues (reviewed by Rueckert 1990). There are five genera of picornaviruses, these are: *Rhinovirus*, *Enterovirus* (includes poliovirus), *Cardiovirus* (includes encephalomyocarditis virus (EMCV), *Aphthovirus* (foot-and-mouth disease virus (FMDV) and Hepatovirus. In each case the viral genomes consist of a single-stranded, positive sense, RNA molecule of between 7.4 and 8.3 kb. The genome is linked to a viral encoded peptide, VPg, at its 5' terminal but most of the RNA within an infected cell lacks this feature and simply has a terminal phosphate group. The RNA acts as a mRNA but also has to act as the template for negative strand RNA synthesis to allow RNA replication and the production of positive sense RNA. Many of the picornaviruses cause a dramatic alteration in the protein synthetic profile of infected cells. This is a result of new

[1] Dept of Biochemistry, McGill University, Montreal, Quebec, Canada H3G IY6
[2] McGill Cancer Center, McGill University, Montreal, Quebec, Canada H3G IY6
[3] AFRC Institute for Animal Health, Purbright, Woking, Surrey, GU24 ONF, UK
[4] Institute of Poliomyelitis and Viral Encephalitides, Russian Academy of Medical Sciences, Moscow Region 142782, Russia

synthesis of viral proteins and also the inhibition of host protein synthesis. The enteroviruses, rhinoviruses and aphthoviruses inhibit host cell protein synthesis as a result of the cleavage of the p220 component of the cap-binding complex (eIF-4F) (ETCHISON et al. 1982; DEVANEY et al. 1988) and possibly by other uncharacterized mechanisms. The cardioviruses also inhibit host protein synthesis but do not induce p220 cleavage (MOSENKIS et al. 1985). The viral protein synthesis is cap-independent.

The initiation of protein synthesis on eukaryotic mRNAs is a complex process and is the target of regulation by several different mechanisms. The vast majority of eukaryotic mRNAs possess a cap structure at their 5' terminals which is required for efficient recognition by a complex of protein synthesis initiation factors (including eIF-4F) and hence for translation. On most mRNAs protein synthesis begins at the AUG codon closest to the 5' terminal which usually resides within 50–100 residues (reviewed by KOZAK 1989). However, in some cases alternate AUG codons are used and these exceptions allowed the derivation of a consensus sequence for efficient initiation. Certain elements upstream of the initiation codon can affect the efficiency of initiation. Secondary structure is inhibitory. The presence of AUG codons between the 5' terminal and the authentic initiation site can also inhibit translation.

Against this background the genomes and RNAs produced by picornaviruses would be expected to be very poorly translated. As already indicated these RNAs lack the cap structure. The initiation codon is positioned some 650–1300 bases (depending on the virus) downstream from the 5' terminal. Furthermore, these long 5' noncoding regions (NCR) are predicted to form extensive secondary structure and contain multiple AUG codons (reviewed by JACKSON et al. 1990, MEEROVITCH and SONENBERG 1993). Despite these features the RNAs produced by EMCV and FMDV are extremely efficient mRNAs when translated in rabbit reticulocyte lysate and replicate and translate very efficiently within cells. Poliovirus (PV) has been the prototype picornavirus but despite its efficient growth in cells its RNA translates rather inefficiently in rabbit reticulocyte lysate. However, in HeLa cell extracts or in rabbit reticulocyte lysate supplemented with HeLa cell lysate the translation occurs much more efficiently (BROWN and EHRENFELD 1979; DORNER et al. 1984). This observation is important since it suggests that cellular factors, in addition to those required for classical translation, are necessary for efficient PV RNA translation.

2 Evidence for Internal Initiation of Protein Synthesis

The information summarized above indicates that a novel mechanism of initiation of translation is used by picornaviruses. The key experiment to elucidate the nature of this mechanism was the construction of plasmids which express artificial bicistronic mRNAs (PELLETIER and SONENBERG 1988; JANG et al. 1988). In eukaryotic systems the first open reading of such an mRNA is efficiently expressed but the second open reading frame is translated at around 1% of this efficiency (see, e.g., NICHOLSON et al. 1991, DREW and BELSHAM 1994) presumably by a low level of reinitiation. The introduction of a picornavirus 5' NCR in the correct orientation, between the two open reading frames, led to efficient expression of the second open reading frame as well (Fig 1). It is now clear from many studies that the translation of the second reading frame directed by a picornavirus 5' NCR is totally independent of initiation of the first open reading frame (eg PELLETIER and SONENBERG 1989, JANG et al. 1989, BELSHAM and BRANGWYN 1990). Suppression of cap-dependent translation inhibits translation of the first open reading frame but the cap-independent translation directed by the picornavirus 5' NCR is unaffected. This is consistent with the fact that many picornaviruses inhibit host cell cap-dependent protein synthesis. The initial identification of internal initiation of protein synthesis directed by the 5' NCR of PV and EMCV has been followed by the identification of functionally analogous regions from representatives of each of the picornavirus genera (BELSHAM and BRANGWYN 1990; KUHN et al. 1990; BROWN et al. 1991; BORMAN and JACKSON 1992). It has become apparent that the five genera include three basic types of element, each having the ability to direct internal initiation. The elements from the rhinoviruses and enteroviruses are similar to each other, as are the elements from the cardioviruses and FMDV, but these two types of element are quite distinct from each other. Hepatitis A has an analogous element which has some similarity to the cardiovirus element but is significantly different (BROWN et al. 1991). The rest of this review will concentrate on studies on the PV 5' NCR. Other chapters in this book will discuss the analogous elements from other picornaviruses. The element which is

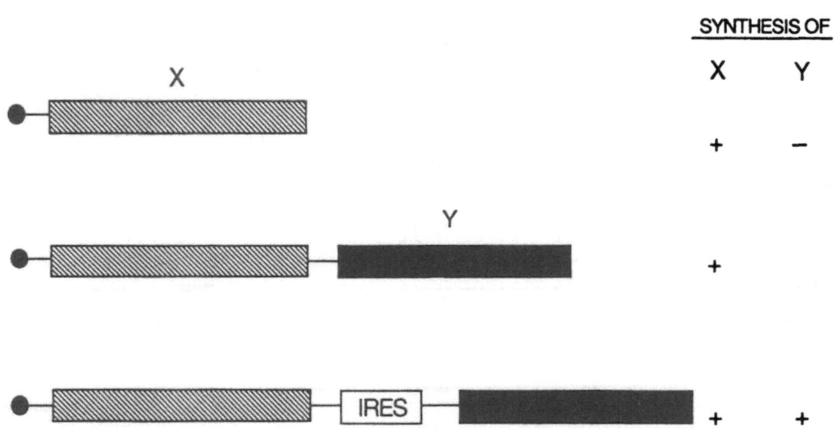

Fig. 1. Principle of the assay for internal initiation of protein synthesis using bicistronic mRNAs. mRNAs containing open reading frames encoding protein X or X and Y as indicated can be translated in vitro or in vivo. The results of such expression studies are indicated. The presence of the IRES directs the synthesis of the protein Y independently of the expression of X. The 5' terminal of the mRNA has a cap structure (indicated by the *circle*) which facilitates recognition of the RNA by eIF-4F, when the p220 component of this complex is cleaved (see text). Translation of X is essentially abolished but the IRES directed translation of Y continues

responsible for directing internal initiation of protein synthesis will be called here the internal ribosome entry site (IRES) since this term has gained widespread usage for all the picornaviruses although previously the term "ribosome landing pad" has been used in this laboratory for the PV element (PELLETIER and SONENBERG 1988).

3 Definition of the Poliovirus Internal Ribosome Entry Site

The 5' NCR of PV is approximately 745 bases in length (Fig. 2). An early task was the identification of the region within this sequence which is necessary and sufficient for IRES function. Deletion analysis from the 5' and 3' terminals was performed and the residual sequences were assayed within monocistronic mRNAs by in vitro translation (PELLETIER et al. 1988a). This assay suggested minimum limits for the IRES of 320–631, and it was shown that the region between 140–320 enhanced the activity. More recent studies, using assays based on the analysis of bicistronic mRNAs in vivo, have modified this slightly and

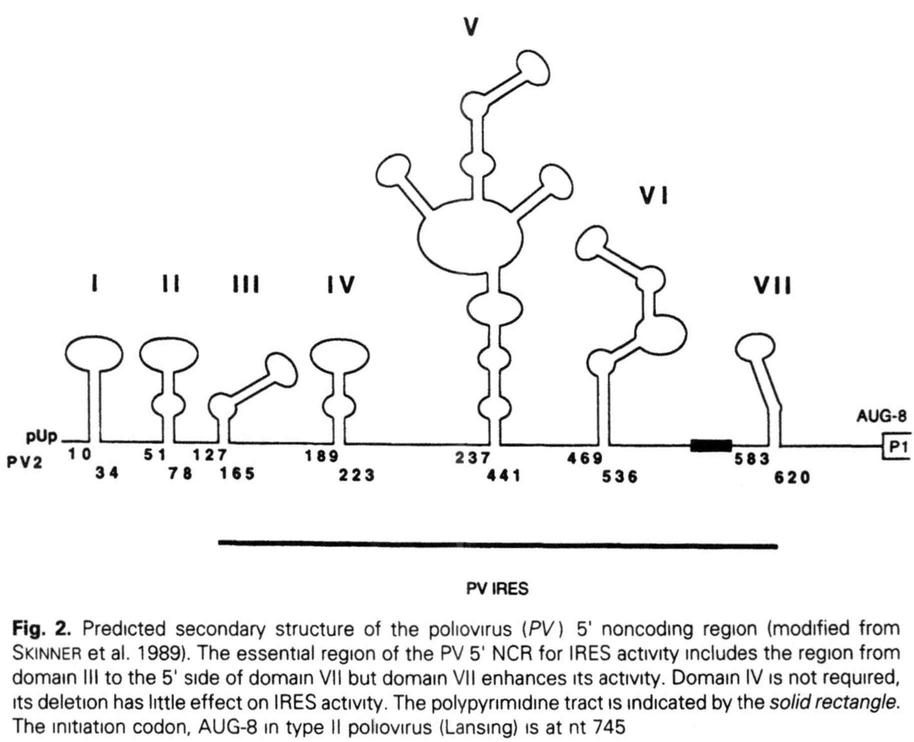

Fig. 2. Predicted secondary structure of the poliovirus (*PV*) 5' noncoding region (modified from SKINNER et al. 1989). The essential region of the PV 5' NCR for IRES activity includes the region from domain III to the 5' side of domain VII but domain VII enhances its activity. Domain IV is not required, its deletion has little effect on IRES activity. The polypyrimidine tract is indicated by the *solid rectangle*. The initiation codon, AUG-8 in type II poliovirus (Lansing) is at nt 745

a consensus (NICHOLSON et al. 1991; PERCY et al. 1992) seems to suggest that the region from 127–620 is required for optimal IRES activity (this includes domains III–VII). The domain VII (583–620) is not essential for IRES function but certainly enhances its activity. Further deletion analysis has shown that an internal region, domain IV (189–223), can be deleted with little effect on IRES function (NICHOLOSON et al. 1991; PERCY et al. 1992; STONE et al. 1993; HALLER et al. 1993). The dispensability of domains IV and VII is in accord with the fact that viable PVs lacking these elements have been isolated (IIZUKA et al. 1989; PILIPENKO et al. 1992; HALLER et al. 1993). To some extent trying to define precise boundaries for the IRES may be unhelpful since the activity of a functional core may be dependent on its freedom to adopt its native conformation. Domain VII is just to the 3' side of a conserved polypyrimidine tract (see below) and contains an AUG codon (AUG-7 in PV type II Lansing) which alone amongst the upstream AUG codons confers an advantage to the virus (PELLETIER et al. 1988b). Mutagenesis of this codon produces a virus with a small plaque phenotype. Modifications of this sequence in type I PV also adversely affected the virus and isolation of revertant viruses showed that the virus selected for an arrangement in which the polypyrimidine tract was followed by an AUG codon about 22 bases downstream (PILIPENKO et al. 1992). It is important to stress that this AUG codon is not the initiator codon which is found ~150 bases downsteam. Hence the PV IRES is separated from the initiator AUG by a considerable spacer. In rhinoviruses much of this spacer is missing. Evidence suggests that the initiation complex "scan" through this spacer region, since inclusion of a 72 base insertion is not deleterious unless it contains an AUG codon (KUGE et al. 1989). In the cardioviruses and aphthoviruses the 3' terminal of the IRES is adjacent to the initiator AUG, although in FMDV a special case exists since a second AUG codon is used as an initiator codon which is some 84 bases further downstream. Evidence indicates that this region is also "scanned" with only a single ribosome entry point (BELSHAM 1992).

4 Structure of the Internal Ribosome Entry Site

The region defined as the PV IRES is about 450 bases in length, similar in size to that defined in EMCV and FMDV. Each of these elements is predicted to contain extensive secondary structure. Biochemical support for the predicted secondary structures has been obtained by probing with RNAses and reagents which have specificity for single-stranded or double-stranded regions of RNA. Such studies, together with phylogenetic comparisons, have been used to derive the structure for the PV 5' NCR shown in Fig. 2 (see SKINNER et al. 1989). Other enteroviruses and rhinoviruses which differ significantly in sequence can be predicted to fold in this manner as a result of compensatory base substitutions. The structure predicted for the cardioviruses and FMDV is totally distinct (PILIPENKO et al. 1989).

Unfortunately in neither case is there much evidence for tertiary interactions within the IRES.

The question therefore arises as to how these diverse elements achieve the same function. As mentioned in the introduction the RNAs of EMCV and FMDV translate very efficiently in rabbit reticulocyte lysate whereas the PV IRES functions relatively poorly in this system unless it is supplemented with HeLa cell proteins, thus suggesting a role for proteins other than the well defined initiation factors in the initiation of protein synthesis directed by the PV IRES. The role of the canonical initiation factors in internal initiation of protein synthesis is not yet well defined. Although it is clear that intact eIF-4F is not required it is possible that the cleavage products of p220 produced in picornavirus infected cells may still be functional. Recent studies using dominant negative mutants of eIF-4A have shown that these inhibit IRES directed translation as well as cap-dependent translation (PAUSE et al. 1994) indicating an essential role for this initiation factor.

Another important finding was the demonstration that key attenuating mutations in PV occurred within the region of 472–481 (WESTROP et al. 1987; NOMOTO et al. 1987; REN et al. 1991). This region is located within domain VI of the IRES. It has been demonstrated that such attenuated viruses replicate poorly in neuroblastoma cells (LA MONICA and RACANIELLO 1989) and their RNA is translated less efficiently than the wild-type RNA in Krebs-2 ascites cell extracts (SVITKIN et al. 1985, 1990). The attenuated viruses still replicate efficiently in HeLa cells. Recognition of the IRES by cellular proteins may be critical to the tropism of the virus and in the case of PV to its ability to induce damage within the nervous system.

5 Role of the Polypyrimidine Tract in Internal Ribosome Entry Site Function

A polypyrimidine tract close to the 3' terminal of the IRES is a highly conserved feature of all picornaviruses. In type II PV this polypyrimidine tract is 15 bases long and includes a single A residue. Mutagenesis suggested that the 5' terminal region UUUCC is critical for function both in vitro and in vivo since replacement of the two Cs of the three Us with G residues severely diminished the efficiency of internal initiation (NICHOLSON et al. 1991). Similar results have been obtained on studies with both EMCV and FMDV (JANG and WIMMER 1990; KUHN et al. 1990). However, recently some evidence has been obtained which indicates a rather minor role for the polypyrimidine tract in some systems. A large number of mutants which altered the length and sequence of the EMCV tract had rather modest effects on the translational efficiency (in vitro and in vivo) and indeed the change of all the pyrimidines to A residues only decreased translational efficiency by about 30% (KAMINSKI et al. 1994). Similar results have also been obtained with another cardiovirus, Theiler's murine encephalomyocarditis virus (TMEV) (T.D.K. BROWN, personal communication). This may suggest that under conditions in which the IRES elements can function very efficiently the role of the polypyrimidine tract is rather limited. Alternatively it may be that the important role of this motif is to maintain an unstructured region of RNA. Introduction of G residues into the polypyrimidine tract may introduce secondary structure and hence interfere with function.

6 Identification of Proteins Interacting with the Poliovirus Internal Ribosome Entry Site

In order to detect cellular proteins which interact with the IRES two types of assay have been performed. A mobility shift assay was used to show that HeLa cell extracts specifically interact with a RNA transcript of residues 559–620 (MEEROVITCH et al. 1989). This segment of RNA was chosen for study since it contains the polypyrimidine tract and the AUG-7 codon described above as playing an important role in the efficient translation directed by the PV IRES. It was shown that a complex formed between a component of the HeLa cell lysate and the 559–620 probe which was distinct from that generated with other known initiation factors. The protein responsible was mainly located in the S100 fraction of the cell. This interaction was assessed as specific since it was competed by the PV RNA transcript but not by other RNAs. The formation of this complex was readily observed with HeLa cell lysate. In contrast only a weak signal indicative of this complex could be detected in rabbit reticulocyte lysate. Analysis of the HeLa cell lysate complex generated in this assay indicated that the transcript was interacting predominantly with a protein of 52 kDa and hence termed p52. This protein can also be readily detected by UV cross-linking the same transcript to total HeLa cell lysate, RNAse digestion and analysis by SDS-PAGE. An array of different protein species have been detected by this UV cross-linking procedure to bind to regions of the PV 5'NCR in different laboratories. In addition to p52 these include proteins of 57 kDa, 50 kDa, 48 kDa, 38 kDa and others including eIF-2 (PESTOVA et al. 1991; NAJITA and SARNOW 1990; DEL ANGEL et al. 1989; GEBHARD and EHRENFELD 1992). Since the PV 5'NCR has a number of functions to perform the significance of any interaction with a protein to the process of translation has to be established. For example, it has been demonstrated that a p50 species is recognized by domain IV (NAJITA and SARNOW 1990), but the significance of this interaction is far from clear since viable PVs lacking this feature can be isolated (HALLER et al. 1993) and indeed bovine enterovirus lacks this element (EARLE et al. 1988). It is clear that some further evidence for the role of any particular protein in internal initiation must be obtained before much credibility can be assigned to such data alone.

Using the gel shift mobility assay it was possible to purify the 52 kDa protein from HeLa cells (MEEROVITCH et al. 1993). The purified protein was analysed in an RNA binding assay and shown to have a high affinity for the 559–624 transcript with an affinity constant of $4x10^{-9}$ *M*. The availability of purified protein permitted its identification. Microsequencing of a tryptic peptide showed complete identity over 22 amino acids with the human La autoantigen. To confirm the identity of p52 as La, further analyses were performed. An anti-La monoclonal antibody specifically recognized the p52. Furthermore, it recognized and "supershifted" the PV RNA-protein complex observed in the gel mobility shift assay. The La protein (reviewed by TAN 1989) is known as the target for autoimmune recognition in patients with systemic lupus erythematosus and Sjogren's syndrome. It has previously been characterized as an RNA-binding protein and has a characteristic RNA recognition motif (termed an RRM). La is reported to bind to the 3' terminals of RNA polymerase III transcripts and function in the maturation of these molecules (GOTTLIEB and STEITZ 1989). The protein is predominantly localized within the nuclei of cells.

7 Functional Significance of the Interaction Between La and Poliovirus RNA

As mentioned above it has long been known that PV RNA translates inefficiently in rabbit reticulocyte lysate. Furthermore the translation produces proteins distinct from those observed in vivo (BROWN and EHRENFELD 1979; DORNER et al. 1984). These aberrant products were shown to result from multiple initiation events within the 3' region of the RNA (DORNER et al. 1984). By contrast, HeLa cell extracts translate PV RNA with great accuracy and under certain conditions the generation of infectious virus from the de novo synthesized components can be observed (MOLLA et al. 1991). Addition of HeLa cell extract to rabbit reticulocyte lysate stimulates the translation of PV RNA and suppresses the production of the aberrant proteins (BROWN and EHRENFELD 1979; DORNER et al. 1984; PHILIPS and EMMERT 1986; SVITKIN et al. 1985, 1988). An important observation was that addition of purified La, expressed in *E. coli*, was also able to stimulate PV RNA translation in rabbit reticulocyte lysate (MEEROVITCH et al. 1993; SVITKIN et al. 1994), and furthermore it showed the same correction activity as observed with HeLa cell lysate (Fig. 3). Thus, the stimulatory and correction activities detected in HeLa cell lysate can be achieved by the single polypetide La. It is noteworthy that a truncated La protein (residues 1–194 out of 408 in the wild-type molecule), which still contains the RRM and binds to the PV RNA, fails to stimulate PV translation or to correct the production of aberrant products (SVITKIN et al. 1994). Hence, merely binding of the protein to the RNA is insufficient to explain its activity. Other proteins, eg. GEF and eIF-2, can also stimulate PV RNA translation within rabbit reticulocyte lysate but do not have the correction activity (SVITKIN et al. 1994).

Recent experiments have shown that recombinant La can also stimulate the translation from the PV IRES within the context of a bicistronic mRNA (Fig. 4) (unpublished results). The CAT/PV IRES/LUC mRNA was translated in the pre-

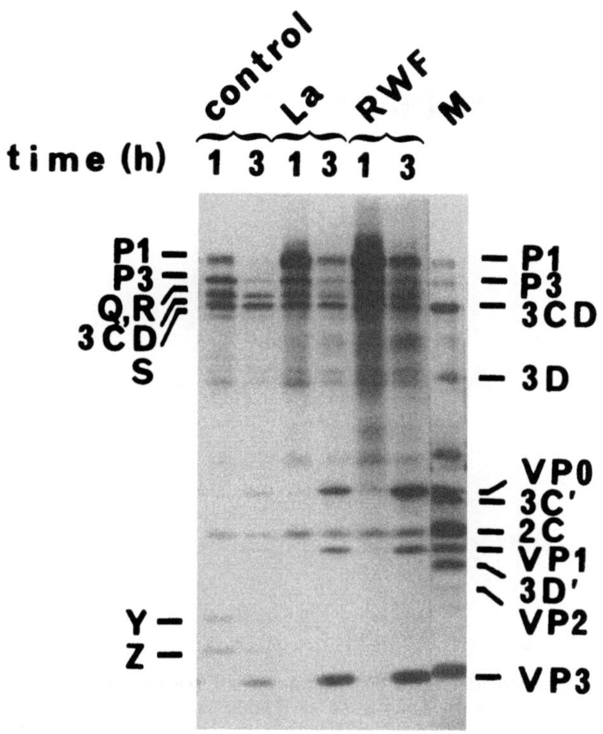

Fig. 3. Stimulation and correction of PV RNA translation in vitro by La. PV RNA was translated in rabbit reticulocyte lysate alone or supplemented with HeLa cell ribosomal salt wash (*RWF*) or with recombinant La protein. The authentic PV protein products (as observed in PV infected cells, lane *M*) and the aberrant initiation products Q, R, S, Y and Z are indicated. Note the increased production of P1 and its cleavage products VP0, VP3 and VP1 in the presence of RWF and La and the reduced synthesis of the aberrant products

sence or absence of La. A marked reduction in the level of CAT synthesis was observed upon addition of La but the production of LUC was stimulated several fold. In the absence of the IRES no stimulation of LUC production was observed. It has been shown previously that La has no effect on the translation of several other viral RNAs (MEEROVITCH et al. 1993). Since all picornaviruses replicate in the cytoplasm and translation occurs in this cell compartment it was surprising that the PV RNA should interact with a protein which is predominantly localized to the cell nucleus. However, it has recently been demonstrated that following PV infection a marked redistribution of La occurs so that La is readily observed within the cytoplasm (MEEROVITCH et al. 1993). It is of course necessary that some cytoplasmic La must be present in uninfected cells to allow the initial cycles of translation to occur upon the initial infection. Indeed, some staining of the cytoplasm of uninfected cells by anti-La antibodies has been reported (HABETS et al. 1983). A prediction from these results is that the depletion of La from HeLa cell

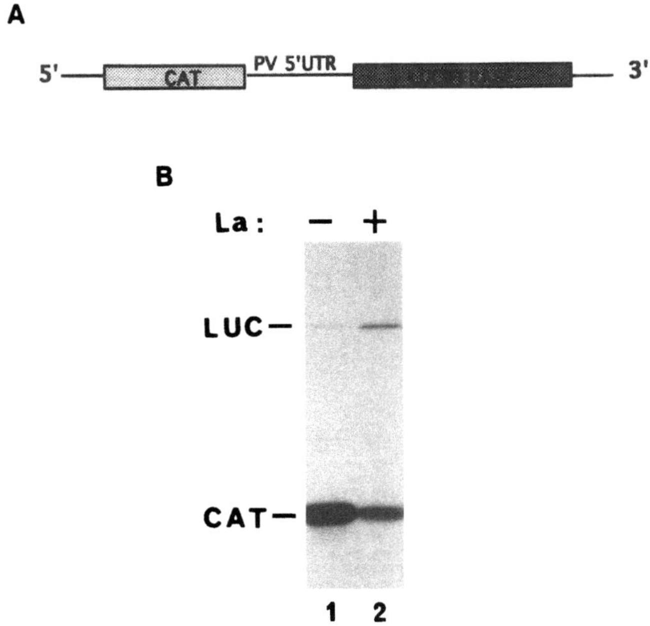

Fig. 4A,B. Stimulation of PV IRES directed translation by La from an artificial bicistronic mRNA. **A** bicistronic mRNA encoding chloramphenicol acetyl transferase (CAT) and luciferase (LUC) separated by the PV 5' NCR was transcribed from pGEMCAT/PV/LUC and **B** translated in rabbit reticylocyte lysate alone or with recombinant La protein. The stimulation of LUC synthesis was about fivefold

lysate should inhibit PV RNA translation. This experiment has been attempted using antibodies specific for La and indeed inhibition of PV RNA translation occurred (YVS, unpublished results). However, it was not possible to restore translation by the addition of purified La. This could indicate that the immunodepletion procedure is insufficiently specific. However, a second possibility is that La is complexed to other proteins and hence the immunodepletion of La also depleted other necessary factors. Similarly, immunodepletion of the p57 (polypyrimidine tract binding protein, PTB) from a HeLa cell lysate inhibited the translation of both PV RNA and EMCV RNA, but again no restoration of translation was achieved with the addition of PTB (Hellen et al. 1993). The major species of protein which cross-links to the EMCV and FMDV IRES is p57 (Jang and Wimmer 1990; Kuhn et al. 1990). The recent identification of p57 as PTB (Borman et al. 1993; Hellen et al. 1993) was again unexpected since this protein has shown to be a factor implicated in RNA splicing within the cell nucleus. Evidence for cytoplasmic location has also been obtained, however. To date, no evidence for a function of the interaction of p57 with the PV RNA has been reported. In EMCV, mutations in the 5' terminal motif of the EMCV IRES (stem-loop E), were reported to inhibit both PTB binding and IRES function (Jang and Wimmer 1990), but these are not necessarily linked and some studies have shown that efficient IRES function can be detected in vitro in the absence of this interaction (Duke et al. 1992).

Fractionation of HeLa cell lysate has also been used to detect proteins which stimulate the activity of the rhinovirus IRES (BORMAN et al. 1993). Column fractions were also tested, using a UV cross-linking assay, for their ability to interact with the rhinovirous IRES. It was shown that the stimulatory activity was separable from the bulk of the p57 (PTB) but did comigrate with a minor fraction of this protein which seemed to be associated with a 97 KDa protein. This species also cross-linked to the rhinovirus 5' UTR. No data was presented on the total composition of the active fractions however, so it is not certain that either the p57 or the p97 were the functional molecules. It was suggested that the two molecules acted synergistically. Purification of the p97 is required before the significance of these data can be assessed.

8 Speculation

It may be that a complex of several cellular proteins (distinct from the standard initiation factors) are required for internal initiation of protein synthesis. Presumably such a complex would also function on the cellular mRNAs which also translate by such a mechanism, e.g. BiP (MACEJAK and SARNOW 1991). An intriguing recent finding is that defective PV IRES elements which lack essential elements can be complemented by the co-expression of an intact homologous IRES element (PERCY et al. 1992; STONE et al. 1993) (analogous results for the complementation of the FMDV IRES have also been obtained (DREW and BELSHAM 1994). An attractive model to explain this phenomenon is that a functional preinitiation complex, presumably containing La and PTB and other proteins, assembles on the intact PV IRES and then transfers to the defective elements, which of necessity include the region around the La binding site (PERCY et al. 1992), and hence translation can be initiated. It may be that such a transfer of a preinitiation complex is a normal part of internal initiation, perhaps partially analogous to the ribosome shunt mechanism proposed for cauliflower mosaic virus 35S RNA translation (FUTTERER et al. 1993). The transfer may normally happen in *cis* but the option to transfer in *trans* may occur when a high concentration of RNA is present or when initiation of translation occurs in a complex with multiple RNAs in close proximity.

References

Belsham GJ (1992) Dual initiation sites of protein synthesis on foot-and-mouth disease virus RNA are selected following internal entry and scanning of ribosomes in vivo. EMBO J 11: 1105–1110

Belsham GJ, Brangwyn JK (1990) A region of the 5' noncoding region of foot-and-mouth disease virus RNA directs efficient internal initiation of protein synthesis within cells: involvement with the role of L protease in translation control. J Virol 64: 5389–5395

Borman AM, Jackson RJ (1992) Initiation of translation of human rhinovirus RNA: mapping the internal ribosome entry site. Virology 188: 685–696

Borman A, Howell MT, Patton JG, Jackson RJ (1993) The involvement of a spliceosome component in internal initiation of human rhinovirus RNA translation. J Gen Virol 174: 1775–1788

Brown BA, Ehrenfeld E (1979) Translation of poliovirus RNA in vitro: changes in cleavage pattern and initiation sites by ribosomal salt wash. Virology 97: 396–405

Brown EA, Day SP, Jansen RW, Lemon SM (1991) The 5' nontranslated region of hepatitis A virus RNA: secondary structure and elements required for translation in vitro. J Virol 65: 5828–5838

Del Angel RM, Papavassiliou Ag, Fernandez-Tomas C, Silverstein SJ, Racaniello VR (1989) Cell proteins bind to multiple sites within the 5' untranslated region of poliovirus RNA. Proc Natl Acad Sci USA 86: 8299–8303

Devaney MA, Vakharia VN, Lloyd RE, Ehrenfeld E, Grubman MJ (1988) Leader protein of foot-and-mouth disease virus is required for cleavage of the p220 component of the cap binding complex. J Virol 62: 4407–4409

Dorner AJ, Semler BL, Jackson RJ, Hanecak R, Duprey E, Wimmer E (1984) In vitro translation of poliovirus RNA: utilization of internal initiation sites in reticulocyte lysate. J Virol 50: 507–514

Drew J, Belsham GJ (1994) trans Complementation by RNA of defective foot-and-mouth disease virus internal ribosome entry site elements J Virol 68: 697–703

Duke GM, Hoffman M, Palmenberg AC (1992) Sequence and structural elements that contribute to efficient encephalomyocarditis virus RNA translation. J Virol 66: 1602–1609

Earle JAP, Skuce RA, Fleming CS, Hoey EM, Martin SJ (1988) The complete nucleotide sequence of a bovine enterovirus. J Gen Virol 69: 253–263

Etchison D, Milburn S, Edery I, Sonenberg N, Hershey JWB (1982) Inhibition of HeLa cell protein synthesis following poliovirus infection correlates with the proteolysis of a 220,000 dalton polypeptide associated with eukaryotic initiation factor 3 and a capbinding complex. J Biol Chem 257: 14806–14810

Fütterer J, Kiss-Laszlo Z, Hohn T (1993) Nonlinear ribosome migration on cauliflower mosaic virus 35S RNA. Cell 73: 789–802

Gebhard JR, Ehrenfeld E (1992) Specific interactions of HeLa cell proteins with proposed translation domains of the poliovirus 5' noncoding region. J Virol 66: 3101–3109

Gottlieb E, Steitz JA (1989) Function of the mammalian La protein: evidence for its action in transcription termination by RNA polymerase III. EMBO J 8: 851–861

Habets WJ, den Brok JH, Boerbooms AMT, van de putte LBA, van Venrooij WJ (1983) Characterization of the SS-B (La) antigen in adenovirus-infected and uninfected HeLa cells. EMBO J 2: 1625–1631

Haller AA, Nguyen JHC, Semler BL (1993) Minimum internal ribosome entry site required for poliovirus infectivity. J Virol 67: 7461–7471

Hellen CUT, Witherell GW, Schmid M, Shin SH, Pestova TV, Gil A, Wimmer E (1993) A cytoplasmic 57-kDa protein that is required for translation of picornavirus RNA by internal ribosomal entry is identical to the nuclear pyrimidine tract-binding protein. Proc Natl Acad Sci USA 90: 7642–7646

Iizuka N, Kohara M, Hagino Yamagishi K, Abe S, Komatsu T, Tago K, Arita M, Nomoto A (1989) Construction of less neurovirulent polioviruses by introducing deletions into the 5' noncoding region sequence of the genome. J Virol 63: 5354–5363

Jackson RJ, Howell MT, Kaminski A (1990) The novel mechanism of initiation of picornavirus RNA translation. Trends Biochem Sci 15: 477–483

Jang S-K, Wimmer E (1990) Cap-independent translation of encephalomyocarditis virus RNA: structural elements of the internal ribosomal entry site and involvement of a cellular 57-kD RNA binding protein. Genes Dev 4. 1560–1572

Jang SK, Kräusslich HG, Nicklin MJH, Duke GM, Palmenberg AC, Wimmer E (1988) A segment of the 5' non-translated region of encephalomycarditis virus RNA directs internal entry or ribosomes during in vitro translation. J Virol 62: 2636–2643

Jang SK, Davies MV, Kaufman RJ, Wimmer E (1989) Initiation of protein synthesis by internal entry of ribosomes into the 5' nontranslated region of encephalomyocarditis virus RNA in vivo. J Virol 63: 1651–1660

Kaminski A, Belsham GJ, Jackson RJ (1994) Translation of encephalomyocarditis virus RNA: parameters influencing the selection of the internal initiation site. EMBO J 13: 1673–1681

Kozak M (1989) The scanning model for translation: an update. J Cell Biol 108: 229–241

Kuge S, Kawamura N, Nomoto A (1989) Genetic variation occurring on the genome of an in vitro insertion mutant poliovirus type 1. J Virol 63: 1069–1075

Kuhn R, Luz N, Beck E (1990) Functional analysis of the internal initiation site of foot–and-mouth disease virus. J Virol 64: 4625–4631

La Monica N, Racaniello VR (1989) Differences in replication of attenuated and neurovirulent polioviruses in human neuroblastoma cell line SH-SY5Y. J Virol 63: 2357–2360
Macejak DG, Sarnow P (1991) Internal initiation of translation mediated by the 5' leader of a cellular mRNA. Nature 353: 90–94
Meerovitch K, Sonenberg N (1993) Internal initiation of picornavirus RNA translation. Semin Virol 4: 217–227
Meerovitch K, Pelletier J, Sonenberg N (1989) A cellular protein that binds to the 5'-noncoding region of poliovirus RNA: implications for internal translation initiation. Genes Dev 3: 1026–1034
Meerovitch K, Svitkin YV, Lee HS, Lejbkowicz F, Kenan DJ, Chan EKL, Agol VI, Keene JD, Sonenberg N (1993) La autoantigen enhances and corrects aberrant translation of poliovirus RNA in reticulocyte lysate. J Virol 67: 3798–3807
Molla A, Paul AV, Wimmer E (1991) Cell-free, de novo synthesis of poliovirus. Science 254: 1647–1651
Mosenkis J, Daniels-McQueen S, Janovec S, Duncan R, Hershey JWB, Grifo JA, Merrick WC, Thach RE (1985) Shutoff of host translation by encephalomyocarditis virus infection does not involve cleavage of the eukaryotic initiation factor 4F polypeptide that accompanies poliovirus infection. J Virol 54: 643–645
Najita L, Sarnow P (1990) Oxidation-reduction sensitive interaction of a cellular 50 kDa protein with an RNA hairpin in the 5' noncoding region of the poliovirus genome. Proc Natl Acad Sci USA 87: 5846–5850
Nicholson R, Pelletier J, Le S-Y, Sonenberg N (1991) Structural and functional analysis of the ribosome landing pad of poliovirus: in vivo translation studies. J Virol 65: 5886–5894
Nomoto A, Kohara M, Kuge S, Kawamura N, Arita M, Kamatsu T, Abe S, Semler BL, Wimmer E, Itoh H (1987) Study on virulence of poliovirus type I using in vitro modified viruses. In: Brinton MA, Rueckert RR (eds) Positive strand RNA viruses. Liss, New York, pp 437–452
Pause A, Methot N, Svitkin Y, Merrick WC, Sonenberg N (1994) Dominant negative mutants of eIF-4A define a critical role for eIF-4F in cap-dependent and cap-independent initiation of translation. EMBO J 13: 1205–1215
Pelletier J, Sonenberg N (1988) Internal initiation of translation of eukaryotic mRNA directed by a sequence derived from poliovirus RNA. Nature 334: 320–325
Pelletier J, Sonenberg N (1989) Internal binding of eukaryotic ribosomes on poliovirus RNA: translation in HeLa cell extracts. J Virol 63: 441–444
Pelletier J, Kaplan G, Racaniello VR, Sonenberg N (1988a) Cap-independent translation of poliovirus mRNA is conferred by sequence elements within the 5' noncoding region. Mol Cell Biol 8: 1103–1112
Pelletier J, Flynn ME, Kaplan G, Racaniello VR, Sonenberg N (1988b) Mutational analysis of upstream AUG codons of poliovirus RNA. J Virol 62: 4486–4492
Percy N, Belsham GJ, Brangwyn JK, Sullivan M, Stone DM, Almond JW (1992) Intracellular modifications induced by poliovirus reduce the requirement for structural motifs in the 5' noncoding region of the genome involved in internal initiation of protein synthesis. J Virol 66: 1695–1701.
Pestova TV, Hellen CUT, Wimmer E (1991) Translation of poliovirus RNA: role of an essential cis- acting oligopyrimidine element within the 5' nontranslated region and involvement of a cellular 57-kilodalton protein. J Virol 65: 6194–6204
Phillips BA, Emmert A (1986) Modulation of the expression of poliovirus proteins in reticulocyte lysates. Virology 148: 255–267
Pilipenko EV, Blinov VM, Chernov BK, Dmitrieva, Agol VI (1989) Conservation of the secondary structure elements of the 5'-untranslated region of cardio- and aphthovirus RNAs. Nucleic Acids Res 17: 5701–5711
Pilipenko EV, Gmyl AP, Maslova SV, Svitkin YV, Sinyakov AN, Agol VI (1992) Prokaryotic-like cis elements in the cap-independent internal initiation of translation on picornavirus RNA. Cell 68: 119–131
Ren R, Moss EG, Racaniello VR (1991) Identification of two determinants that attenuate vaccine-related type-2 poliovirus. J Virol 65: 1377–1382
Rueckert RR (1990) Picornaviridae and their replication. In: Fields BN, Knipe DM, Chanock RM, Hirsch MS, Melnick JL, Monath TP, Roizman B, Virology, 2nd edn. Raven, New York, pp 507–548
Skinner MA, Racaniello VR, Dunn G, Cooper J, Minor PD, Almond JW (1989) New model for the secondary structure of the 5' noncoding RNA of poliovirus is supported by biochemical and genetic data that also show that RNA secondary structure is important in neurovirulence. J Mol Biol 207: 379–392
Stone DM, Almond JW, Brangwyn JK, Belsham GJ (1993) trans Complementation of cap-independent translation directed by poliovirus 5' noncoding region deletion mutants: evidence for RNA-RNA interactions. J Virol 67: 6215–6223

Svitkin YV, Maslova SV, Agol VI (1985) The genomes of attenuated and virulent poliovirus strains differ in their in vitro translation efficiencies, Virology 147: 243–252

Svitkin YV, Pestova TV, Maslova SV, Agol VI (1988) Point mutations modify the response of poliovirus RNA to a translation initiation factor. a comparison of neurovirulent and attenuated strains. Virology 166: 394–404

Svitkin YV, Cammack N, Minor PD, Almond JW (1990) Translation deficiency of the Sabin type 3 poliovaccine genome: association with the attenuating mutation C472–> U. Virology 175: 103–109

Svitkin YV, Meerovitch K, Lee HS, Dholakia JN, Kenan DJ, Agol VI, Sonenberg N (1994) Internal translation initiation on poliovirus RNA: further characterization of La function in poliovirus translation in vitro. J Virol 68: 1544–1550

Tan EM (1989) Antinuclear antibodies: diagnostic markers for autoimmune diseases and probes for cell biology. Adv Immunol 44: 93–151

Westrop GD, Evans DMA, Minor PD, Magrath D, Schild GC, Almond JW (1987) Investigation of the molecular basis of attenuation in the Sabin type 3 vaccine using novel recombinant polioviruses constructed from infectious cDNA. In. Rowlands DJ, Mayo MA, Mahy BWJ (eds) The molecular biology of the positive strand RNA viruses. Liss, New York, pp 53–60

Structure and Function of the Hepatitis C Virus Internal Ribosome Entry Site

C. Wang[1,2] and A. Siddiqui[1]

1 Introduction

Human hepatitis C virus (HCV) infects hepatocytes and the viral infection often develops into chronic disease, liver cirrhosis and hepatocellular carcinoma (Saito et al. 1990; Plagemann 1991; Ruiz et al. 1992). Currently HCV infections account for about 80% of posttransfusion hepatitis worldwide. To date HCV has not been successfully grown in any cultured cells (Shimizu et al. 1993) which precludes the biological studies of this virus. The discovery of the human HCV stems from an unprecedented use of molecular cloning techniques rather than by the conventional method of virus isolation (Choo et al. 1989). Nucleic acids extracted from the sera of chimpanzees infected with infectious non-A, non-B hepatitis plasma were used

[1] Department of Microbiology, Program in Molecular Biology, University of Colorado Health Sciences Center, 4200 East Ninth Avenue, Denver, CO 80262, USA
[2] Present address: Department of Cell Biology, Harvard Medical School, 25 Shattuck Street, Boston, MA 02115, USA

to prepare a λ gt11 library which was then immunoscreened with non-A, non-B hepatitis serum (Choo et al. 1989). The viral RNA genome cloned as a cDNA was characterized by nucleotide sequence determination to uncover the genetic order of the coding potential. Further comparison of the complete HCV genome to other known viral genomes placed this virus in the family of Flaviviridae (Miller and Purcell 1990; Houghton et al. 1991). Members of this virus group include human flaviviruses (e.g., Yellow fever virus, Dengue virus and St. Louis encephalitis virus) and animal pestiviruses (e.g., bovine viral diarrhea virus and hog cholera virus).

The viral genome of HCV is represented as a single-stranded RNA molecule with positive polarity, composed of about 9400 nucleotides (Kato et al. 1990; Choo et al. 1991; Inchauspe et al. 1991; Okamoto et al. 1991; Takamizawa et al. 1991). The viral RNA contains a single large open reading frame (ORF) encoding a polyprotein of 3010–3033 amino acids that is subsequently processed into individual viral proteins by both viral and cellular proteases (Fig. 1) (Grakoui et al. 1993; Hijikata et al. 1993; Selby et al. 1993). The viral RNA genome contains a relatively long 5' noncoding region (5'NCR) and a short 3' NCR with a possible poly-A or poly-U tail (Han et al. 1991; Han and Houghton 1992). While remarkable nucleotide diversity exists among HCV subgroups, the 5' NCR maintains a very high degree of nucleotide conservation (>93%) (Okamoto et al. 1991). The 5' NCR varies in length from 332 to 341 nucleotides depending upon the subtype (Han et al. 1991; Takamizawa et al. 1991; Chen et al. 1992; Tanaka et al. 1992). From the phylogenetic sequence comparison, HCV appears to be more closely related to pestiviruses than flaviviruses. For instance, both HCV and pestiviruses contain a relatively long 5' NCR with several AUG codons, whereas flaviviruses contain a short 5' NCR of about 95–132 nucleotides in length without an AUG (except tick-borne encephalitis virus) (Rice et al. 1986; Collett et al. 1989a; Pletnev et al. 1990). A type 1 cap structure ($m^7GpppAmp$) has been found at the 5' end of flavivirus RNA (Chambers et al. 1990) which may have resulted from the action of the

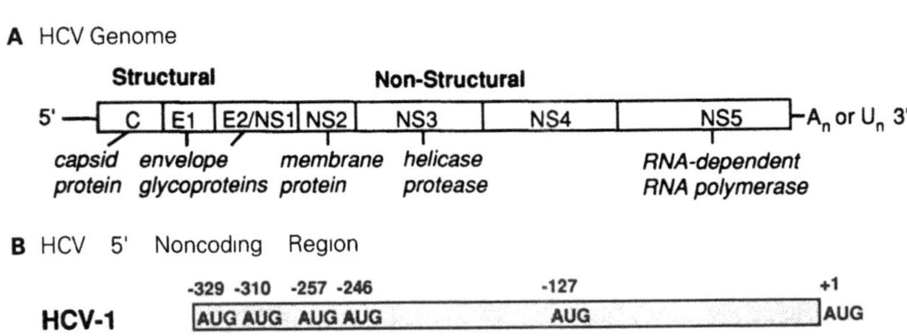

Fig. 1. A Genome organization of the hepatitis C virus (*HCV*) RNA and polyprotein processing scheme. **B** The 5' noncoding region of HCV-1 subtype. Location of various AUG codons in the region is indicated

proposed methyltransferase activity in the flaviviral NS5 protein (KOONIN 1993). The translation of flaviviral RNAs is mediated by a cap-dependent scanning mechanism, and usually initiated at the 5' end proximal AUG triplet (CHAMBERS et al. 1990). In contrast to flaviviruses, multiple AUG triplets have been found in the long 5' NCR of HCV and pestiviruses. The methyltransferase gene found in the flavivirus NS5 region has not been found in the HCV genome. Several experimental approaches have failed to detect a cap structure at the 5' end of pestivirus genomic RNA (COLLETT et al. 1989b; BROCK et al. 1992). It is not known whether 5' end of HCV RNA contains a cap structure. By analogy to pestiviruses, it is likely that the HCV genomic RNA is uncapped. While several AUG codons are located upstream of the translation initiation site of the HCV polyprotein, these do not appear to function as alternative start sites (TSUKIYAMA-KOHARA et al. 1992). Additionally, a complex secondary structure has been proposed for the 5' NCR of the HCV RNA from phylogenetic comparative sequence analysis (BROWN et al. 1992) (Fig.2). This further precludes the possibility of the cap-dependent scanning mechanism for efficient translation of the HCV RNA. These features indicate that translation of HCV RNA may take place by a strategy different from that involving a cap-dependent scanning mechanism (KOZAK 1992).

Picornaviruses symbolize the viral mRNAs that utilize a cap-independent internal ribosome binding mechanism of translation (JACKSON et al. 1990; AGOL

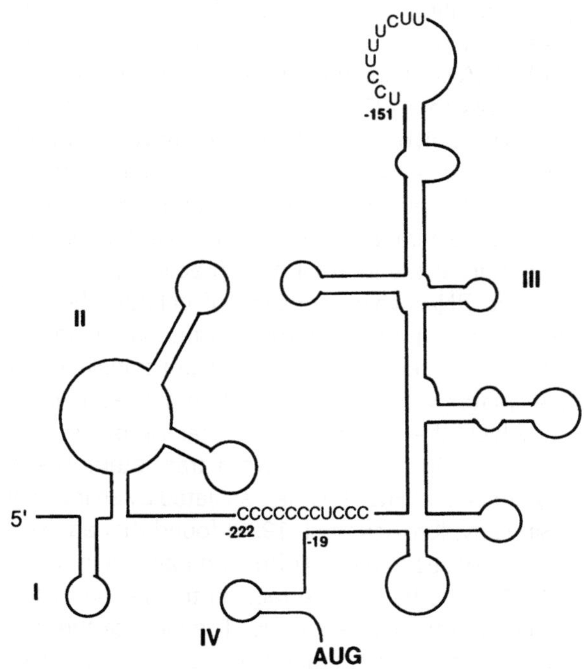

Fig. 2. Map of the computer predicted RNA secondary structure of the 5'NCR of HCV RNA (adapted from BROWN et al. 1992). The nucleotide numbered as +1 position refers to the first A residue of the initiator AUG. The sequences of the two pyrimidine tracts and its locations are shown

1991). This mechanism involves binding of the ribosome to a RNA sequence that has been termed the "internal ribosome entry site" (IRES) (JANG et al. 1989) or "ribosome landing pad" (RLP) (PELLETIER and SONENBERG 1988) within the 5' NCR of the viral RNAs. The picornavirus RNAs contain an unusually long 5' NCR (600 to over 1,000 nucleotides) with multiple AUG triplets and the viral RNAs lack a m^7G cap structure found in all eukaryotic mRNAs. In this respect, the 5' NCRs of HCV and pestiviruses resemble those of picornaviruses. These characteristics have prompted investigations into the possibility of translation of HCV and pestivirus RNAs by a mechanism similar to that employed by picornaviruses. In this review, we attempt to summarize recent work on the translational regulation of HCV and pestivirus RNAs and draw comparisons with those of picornaviruses.

2 Translation of Hepatitis C Virus RNA Is Mediated by Internal Ribosome Entry

2.1 Hepatitis C Virus RNA Contains an Internal Ribosome Entry Site Within Its 5' NCR

The first line of evidence in demonstrating the presence of an IRES element within the 5' NCR of HCV comes from in vitro translation studies by TSUKIYAMA-KOHARA et al. (1992). Using an HCV RNA containing the 5' NCR and some coding sequences of the polyprotein (nt –324 to +1772, +1 position refers to the A residue of the initiator AUG of the viral polyprotein), they have shown that both capped and uncapped HCV RNAs were active as mRNAs with similar translational efficiencies in rabbit reticulocyte lysates. The translation products from the HCV RNA were confirmed as natural viral polypeptides using anti-HCV antibodies. Translation was shown to start at the internal AUG preceding the large open reading frame. In a dicistronic RNA which consists of chloramphenicol acetyl transferase (CAT) reporter gene as the first cistron and the partial coding sequences of the HCV polyprotein as the second cistron, efficient translation of both genes in rabbit reticulocyte lysates was demonstrated. In coxsackievirus-infected HeLa cell extracts, in which the cap-dependent translation is suppressed, only the hepatitis C viral coding sequences were translated. These results provided the first direct evidence that HCV RNA is translated by an internal ribosome binding mechanism. Similarly, KETTINEN et al. 1993) found that addition of cap analogue in in vitro translation reactions had no effect on translation of the uncapped RNA with the HCV 5'NCR and had little effect on translation of the capped RNA at higher concentrations of the cap analogue. The yield of the first cistron from a dicistronic RNA was significantly increased by capping of the RNA, but expression of the second cistron was unaffected. Both reporter gene products were detectable by SDS-PAGE analysis at the earliest time point of translation of the dicistronic RNA, indicating simultaneous initiation of translation of the two genes. These observations together suggest that translation of the HCV RNA is most probably cap-independent.

In support of the in vitro analyses of the translational regulation of the HCV RNA, experiments carried out in this laboratory (WANG et al. 1993) have employed an in vivo system by introducing in vitro synthesized RNA into HepG2 cells, a hepatoblastoma cell line. The 5' NCR of HCV was inserted between two reporter genes, those for CAT and luciferase (LUC). The expression of the LUC gene from these RNAs is dictated by the HCV 5' NCR. Since the in vitro synthesized RNAs were uncapped, the first cistron encoding the CAT protein was poorly translated in transfected cells. CAT activity was not detectable in the cell lysates. However, expression of the second cistron (LUC) occurred when the full-length (nt –341 to –1) or a nearly full-length 5' NCR (nt –304 to –1) were present upstream of the cistron. Since this expression occurs independent of CAT translation, it is unlikely that LUC expression resulted from ribosome reinitiation. These data obtained from in vitro and in vivo experiments provide the unequivocal evidence for the presence of an IRES element within the HCV 5' NCR. RNA transfections of both hepatic and nonhepatocyte-derived cell lines were equally efficient, suggesting a lack of liver cell-specific preference for translation of the HCV RNA.

Recently, in an effort to study the cap-independent translation in yeast, IIZUKA et al. (1994) observed that the HCV 5'NCR is capable of programming cap-independent translation of uncapped monocistronic RNA in yeast cell extract. The efficiency of translation was enhanced more than twofold when the message was polyadenylated. In the presence of cap analogue (m^7GpppG), translation of the RNA was not inhibited, but appeared to be enhanced. However, translation of capped and polyadenylated luciferase mRNA, lacking an IRES element, was inhibited in the presence of the cap analogue. The success of using the yeast system for investigating mechanism of IRES function provides future prospects of identifying factors involved in internal initiation of translation.

Interestingly, hepatitis A virus (HAV), a liver-tropic virus belonging to the picornavirus family, also uses the internal ribosome binding mechanism of translation (GLASS et al. 1993; BROWN et al. 1994). Translation of HAV RNA is inefficient in reticulocyte lysates and results in aberrant initiation (JIA et al. 1991; BROWN et al. 1991). However, addition of cell extracts prepared from fresh mouse liver into the reticulocyte lysates specifically stimulated the cap-independent translation of the HAV RNA (GLASS and SUMMERS 1993). Similar extracts prepared from the kidney, heart or brain of the mouse and from other cultured cells did not show this stimulation. We have not observed any enhancement of the HCV 5'NCR-mediated translation following addition of the cell extracts from cultured HepG2 cells or HeLa cells (WANG and SIDDIQUI, unpublished data). However, it will be interesting to determine whether translation of the HCV RNA could be stimulated by addition of the extracts from the fresh liver tissue.

2.2 Hepatitis C Virus Internal Ribosome Entry Site Element Is Active in Picornavirus-Infected Cells or Extracts

Poliovirus translation occurs in a cap-independent manner (JACKSON et al. 1990; AGOL 1991). The viral infection also results in a shut-off of host protein synthesis. Cleavage of p220, a component of eIF-4F, has been suggested as a possible mechanism which initiates these events (LEE and SONENBERG 1982). We have introduced RNA containing an HCV 5'NCR (T7C1-341) into the poliovirus-infected HepG2 cells at a time when cellular protein synthesis is inhibited (WANG et al. 1993). Translation of the uncapped T7C1-341 RNA occurred efficiently in both uninfected cells and the infected cells (Table 1), whereas translation of normally capped RNA without the HCV 5'NCR (T7-LUC) was dramatically decreased in infected cells. Similar results were also obtained in in vitro translation studies using coxsackievirus-infected cell lysates under conditions of cap-dependent translation inhibition (TSUKIYAMA-KOHARA et al. 1992). These data further support that HCV RNA is translated by a cap-independent mechanism. In this respect, HCV utilizes a translational regulatory mechanism that is similar to that of picornaviruses.

2.3 Mapping of the Hepatitis C Virus Internal Ribosome Entry Site Element

Deletion analysis was used to define the minimal sequences within the 5' NCR that constitute the HCV IRES element. In vivo translation studies by WANG et al. (1993) indicate that the 5' noncoding sequences downstream of nt –304 are essential for efficient translation. We have also found that the sequence at the 3' end of the 5' NCR is critical for IRES function because a short deletion of nine nucleotides at this end abolished the translation. However, deletion of up to nt –257 from the 5' end significantly reduced the efficiency of translation, while the 5' end deletions beyond nt –257 were totally inefficient in initiating translation from an internal site. Similar results were also obtained from in vitro translation studies in rabbit reticulocyte lysates. Deletion mutagenesis of the 5'NCR by TSUKIYAMA-KOHARA et al. (1992) show that deletion of up to nt –231 did not adversely

Table 1. Translation of RNAs in poliovirus-infected and uninfected HepG2 cells

RNA	LUC activity (light units/5×10^5 cells	
	Uninfected	Infected
T7-LUC (capped)	1672	190
T7C1-341 (uncapped)	2453	1208
pPB310 (uncapped)	2626	1081

LUC, luciferase.

affect translation. While this deletion was not tested in vivo, another possible explanation for this discrepancy is the variation of salt concentrations used in in vitro translation experiments. As shown by KETTINEN et al. (1993), translation directed by the HCV 5'NCR was influenced by salt concentrations. Similarly the translation of EMCV (encephalomyocarditis virus) and FMDV (foot and mouth disease virus) were shown to be affected by salt concentrations (JACKSON 1991). KETTINEN et al. (1993) made an interesting observation that coding sequences of the HCV capsid protein may also be involved in HCV IRES function. The data from other laboratories (TSUKIYAMA-KOHARA et al. 1992; WANG et al. 1993) did not show a requirement of the HCV coding sequences for internal initiation of translation. It is, however, possible that the HCV coding sequences could contribute to maximal efficiency of translation initiation by the 5'NCR. Further studies are needed to pursue this interesting finding.

Using site-directed mutagenesis, REYNOLDS et al. (1994), showed that the 5' proximal AUG of the multiple AUG triplets within the 5'NCR is critical for IRES function. This AUG triplet, located in a loop of domain II (Fig. 2), is conserved in all HCV subtypes. Based on mutational analysis it appears that sequences in this domain mark the 5'boundary of the HCV IRES.

3 Hepatitis C Virus Internal Ribosome Entry Site Element

3.1 The Hepatitis C Virus Internal Ribosome Entry Site Element May Not Require a Pyrimidine Tract for Translation Initiation

A conserved feature found in all 5'NCR of picornavirus RNAs that conduct translation by cap-independent mechanism (JACKSON et al. 1990) is the presence of a pyrimidine tract followed by an AUG triplet, the Yn-Xm-AUG motif (WIMMER et al. 1993). Extensive studies on the pyrimidine tract by deletion and point mutations have demonstrated that this motif is essential for picornavirus IRES function (JANG and WIMMER 1990; KUHN et al. 1990; MEEROVITCH et al. 1991; PILIPENKO et al. 1992). The pyrimidine tract must be properly spaced from the AUG triplet for maximal efficiency of translation (PILIPENKO et al. 1992). Scanning the HCV 5'NCR sequences reveals the presence of two putative pyrimidine tracts (Fig. 2). The first pyrimidine tract is located between nt –222 and –212, immediately upstream of the domain III structure. This pyrimidine tract, at the primary sequence level, is not followed by an AUG triplet, an important feature of the Yn-Xm-AUG motif. However, according to the predicted secondary structure of the 5'NCR, a base pairing interaction between part of the pyrimidine tract and a distant nucleotide sequence (nt –217 to – 209 and nt –27 to – 19) brings the pyrimidine tract about 20 nucleotides from the initiator AUG codon (BROWN et al. 1992). The second putative pyrimidine tract is located within the apical loop of the domain III structure (Fig. 2) and includes an AUG triplet about 16 nucleotides downstream. This sequence combination appears to contain the characteristic features of the picornavirus IRES elements (TSUKIYAMA-KOHARA et al. 1992).

A recent study from this laboratory addressed the functional role of the Yn-Xm-AUG motif in HCV IRES function by introducing point mutations in defined regions of the two pyrimidine tracts (WANG et al. 1994). Base substitutions in the nucleotide sequences of the first pyrimidine tract that participate in base pairing proved deleterious for translation initiation. Compensatory mutants which maintain the base substitutions in the pyrimidine tract but restore the helical structure regained efficient translation. One of the compensatory mutants, in which almost half of the pyrimidine bases were altered to purines, did not show any appreciable decrease in translational efficiency compared to the wild-type 5'NCR. Additional mutations in the second pyrimidine tract and the relevant downstream AUG also maintained efficient translation initiation. These results indicate that both pyrimidine tracts are dispensable for HCV IRES function. Our data also support the RNA folding model proposed by BROWN et al. (1992). In a different model, proposed by TSUKIYAMA-KOHARA et al. (1992), the base pairing interaction between nt –217 to – 209 and nt –27 to – 19 was not included. This latter model, therefore is not in agreement with the mutagenesis data.

The mutational analysis of the EMCV 5'NCR recently reported by KAMINSKI et al. (1994) show that complete substitution of the pyrimidines by purines in the pyrimidine-rich tract resulted only in a negligible decrease in translational efficiency. They concluded that the pyrimidine-rich tract was not essential for EMCV IRES function. These findings, combined with the results on HCV, indicate that the primary sequences of the pyrimidine tract may not be critical elements of the IRES. While the pyrimidine tract may have some other functional role in the overall regulation of IRES-mediated translation, its previously envisioned role as an indispensable component of IRES needs to be reevaluated.

3.2 A Possible Pseudoknot Structure Is Critical for Hepatitis C Virus Internal Ribosome Entry Site Function

A closer examination of the nucleotide sequences of the HCV 5' NCR reveals that a possible pseudoknot structure could be formed in the region near the initiator AUG (Fig. 3). The first pyrimidine tract-related helical structure of the HCV 5'NCR, described above, is also included in this putative tertiary structure. Point mutations in each of the complementary strands of the second putative stem within the probable pseudoknot structure abolished translation (Fig. 4) (WANG et al. 1995). When the base pairing interaction was restored with a compensatory mutation, translation was partially recovered. This partial recovery of the translational efficiency from the compensatory mutation may also suggest that the primary sequences in the mutated regions are important. Based on these mutagenesis studies, we propose that a probable pseudoknot structure which also includes the helical structure related to the first pyrimidine tract may be formed. These structural elements together may constitute the critical component of HCV IRES. This tertiary structural folding would also be consistent with RNAse mapping analysis of that region in the 5'NCR (BROWN et al. 1992). However, detailed studies carefully probing into the structures of that region will be needed to substantiate mutational data.

Sequence analysis of all the HCV isolates thus far sequenced show that the regions implicated in secondary and tertiary interactions discussed above are all highly conserved. Similar pseudoknot structures can be predicted in the 5'NCR of the pestivirus RNAs in the same region. Recent studies with pestivirus 5'NCR show the presence of an IRES element (POOLE et al. 1995; also see Sect. 4).

Pseudoknot structures have been implicated in translational regulation (SCHIMMEL 1989; LEATHERS et al. 1993), i.e., in ribosomal frameshifting and translational read-through (BRIERLEY et al. 1989; CHAMORRO et al. 1992; FENG et al. 1992; TZENG et al. 1992; WILLS et al. 1991). In the case of ribosomal frameshifting, a pseudoknot structure functions to stall ribosomes engaged in translating the RNA (BRIERLEY et al. 1991; TU et al. 1992). This pausing increases the efficiency of ribosomal frameshifting. The structural feature in a pseudoknot cannot be

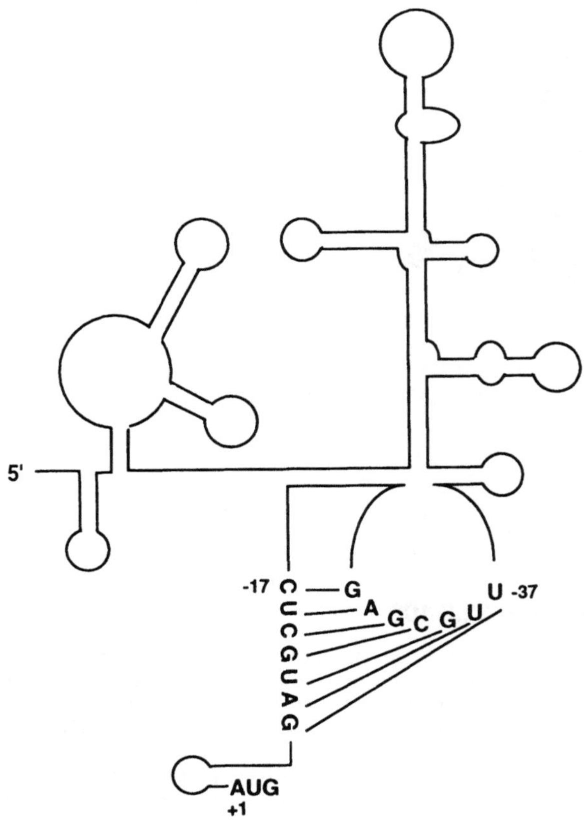

Fig. 3. Predicted pseudoknot structure in the 5'NCR of hepatitis C virus RNA. The possible base pairing interaction between two previously proposed single-stranded regions are shown

replaced by a hairpin structure composed of an equivalent set of base pairs (Brierley et al. 1991; Somogyi et al. 1993). If distinct configurations of a pseudoknot are destroyed, the relevant biological function is affected. Pseudoknot structures have also been implicated to serve as recognition sites for protein binding. For example, a pseudoknot structure in the 5' NCR of bacteriophage T4 gene 32 mRNA is recognized by the protein product of the mRNA (Shamoo et al. 1993), the interaction interfering with further translation of gene 32 mRNA. Thus, it is possible that the pseudoknot structures in the HCV 5' NCR could serve as efficient binding sites for ribosomes and/or relevant initiation factors involved in internal initiation of translation.

Existence of conserved pseudoknot structures have recently been proposed for picornavirus 5'NCR and RNA 3 of infectious bronchitis virus (Le et al. 1992, 1993,1994). The latter is also capable of directing internal ribosome binding (Liu and Inglis 1992). According to this model, the pyrimidine-rich tract of picornaviruses is predicted to be involved in the formation of pseudoknot structures. In light of the proposal, it is likely that mutations in the pyrimidine tract of the 5' NCR of picornaviruses could affect these structures (Jang and Wimmer 1990; Kuhn et al. 1990; Meerovitch et al. 1991; Pilipenko et al. 1992). This provides an alternative explanation for the inhibitory effect of mutations in the pyrimidine-rich sequen-

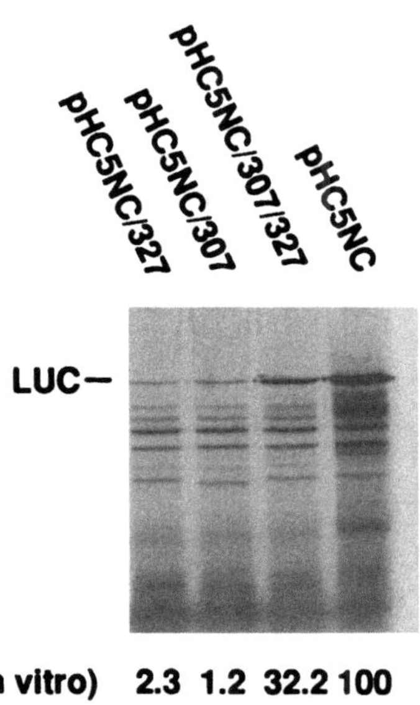

Fig. 4. Mutational studies in one stem of the suspected pseudoknot structure of the HCV 5'NCR. pHC5NC construct consists of a wild-type HCV 5'NCR upstream of the coding sequences of the reporter gene (luciferase). pHC5NC/327 contains nucleotide substitutions at nt –15 – –13 from CGU to GCA; pHC5NC/307 contains the nucleotide substitutions at nt–35––33 from GCG to UGC; and pHC5NC/307/327 is a compensatory mutant containing mutations both at nt –15 – –13 and nt –35 – –33. SDS-PAGE pattern of the ^{35}S-labeled translation products of the RNAs in rabbit reticulocyte lysates is shown. Position of luciferase protein (*LUC*) is marked. The *bottom line* indicates the percentage of luciferase activity from these RNAs in in vitro translation. The luciferase activity is normalized against the value obtained with pHC5NC, which is arbitrarily set at 100

ces, which may have resulted from structural perturbances rather than primary sequence alterations. Further mutational studies in the relevant sequences predicted to be involved in the tertiary structures are needed to substantiate the model.

3.3 Integrity of Domain III Structure in Translation Initiation

According to the RNA folding model of HCV 5'NCR (BROWN et al. 1992), there are at least four structural domains preceding the initiator AUG. Domain III represents the largest secondary structure with multiple stable stem-loops (Fig. 2) and includes a pyrimidine tract within its apical loop. Mutagenesis of this pyrimidine tract, described earlier, indicated that the motif was dispensable for IRES function. This also suggested that the primary sequence in the apical loop of domain III is not important for HCV IRES function. A three base substitution in the stem structure immediately downstream of the apical loop, which is expected to result in broad structural alterations in the upper portion of domain III, also exhibited only a modest effect on translation initiation (WANG et al. 1994). These mutations suggest that the upper portion of the domain III may not be essential for HCV IRES function. A deletion of 127 nucleotides that eliminated a major part of the domain III structure, however, resulted in dramatic reduction of the translational efficiency (WANG et al. 1994). This indicates that integrity of the domain III structure is crucial for translation initiation. Interestingly, in the two groups of picornavirus IRES elements, although similar multiple stem-loop structures are found, their secondary structural arrangements are different (PILIPENKO et al. 1989; SKINNER et al. 1989; BROWN et al. 1991; DUKE et al. 1992). This may imply that certain structural alterations in this domain can also be tolerated. Major deletions in these structural domains of poliovirus and EMCV 5'NCR abrogated translation initiation (NICHOLSON et al. 1991; DUKE et al. 1992; PERCY et al. 1992).

3.4 The 3' Border of the 5'NCR

Picornaviral IRES elements require a spacing of about 20 nuleotides between the AUG and an upstream structural motif for efficient translation (JACKSON et al. 1990; KAMINISKI et al. 1994). In the case of HCV the mutations in the nucleotide sequences upstream of the initiator AUG dramatically affected the efficiency of IRES function. A nine nucleotide deletion mutation was introduced at the 3' border of the HCV 5'NCR upstream of the initiator AUG. This deletion led to the loss of IRES function (WANG et al. 1993). Similarly, the insertion of a random sequence of 32 nucleotides immediately upstream of the initiator AUG also resulted in a significant reduction of translation initiation. Interestingly, in a report by Yoo et al. (1992), 5'NCR constructs that contained an additional 30 nucleotides between the 3'border of the 5'NCR and the AUG of the reporter gene failed to exhibit IRES function. These results suggest that proper spacing between a functional motif of the HCV 5'NCR and the initiator AUG is required for efficient translation. In light of the predicted tertiary structure which involve base pairing interactions of nucleotides in the immediate vicinity of the initiator AUG, proper spacing between these elements might be required for HCV IRES function. Thus, both deletion and insertion mutations may lead to disruption of the spatial relationship between the AUG and the predicted superstructure of the HCV 5'NCR and cause a dramatic reduction of translational efficiency. An alternative explanation is that these mutations may have simply destroyed the proposed superstructure in the vicinity of the initiator AUG with deleterious effects on translation. In any event, these observations implicate the initiator AUG as an essential component of the HCV IRES. In this respect and regarding the ability to efficiently direct internal initiation of translation in vitro in reticulocyte lysates, HCV IRES appears to be similar to the cardiovirus apthovirus group of IRES elements of picornaviruses.

Site-directed mutagenesis studies of the initiator AUG showed that internal initiation was only slightly compromised (a decrease of 10%–20%) when AUG was replaced by AUU or CUG; but if substituted by AAG, GAG, or GCG, the in vitro translated product was slightly smaller in size. This indicates that initiation switched to downstream codons (Reynolds et al. 1995).

4 Translation of the Pestivirus RNA

As discussed earlier, the 5'NCR of the pestiviruses shares several features with the 5'NCR of HCV. There are short regions of conserved primary nucleotide sequences in the 5'NCR that are shared between HCV and the pestiviruses (Bukh et al. 1992). The predicted secondary structures of the 5'NCR of the two groups of viruses display remarkable similarities (Brown et al. 1992; Deng and Brock 1993). Bovine viral diarrhea virus (BVDV), the prototype virus of the pestiviruses, has a 5'NCR of about 385 nucleotides (NADL strain) with six AUG triplets (Collett et al. 1988; Moerlooze et al. 1993). In vitro translation studies have demonstrated that these AUGs are not used as alternative start sites for translation (Wiskerchen et al. 1991). Addition of cap analogue to the in vitro translation reaction had no qualitative or quantitative effect on the translation of BVDV RNA, suggesting a cap-independent mechanism of translation. The use of a translational hybrid arrest assay (Shih et al. 1987), has provided additional evidence in support of the involvement of an internal ribosome binding mechanism in translation of BVDV RNA (Poole et al. 1995). The rationale of this approach is that translation of a mRNA will be normally affected by antisense oligonucleotides complementary to the 5'NCR if translation is directed by the ribosome scanning mechanism. This effect is caused by ribosome arrest in the hybrid regions during the scanning process. However, oilgonucleotides interacting with regions that are outside of the IRES element will have no effect on translation if it occurs through internal ribosome entry. Poole et al. (1994) found that oligonucleotides complementary to the regions between nt 154 and the translation initiator AUG at nt 386 inhibited translation of BVDV RNA, but the oligonucleotides annealed to regions both upstream and downstream of those regions had no effect. This suggests that sequences between nt 154 and the 3'boundary of the BVDV 5'NCR may contain an internal ribosome entry site.

POOLE et al. (1995) have cloned the 5'NCR of BVDV upstream of the second cistron in a dicistronic construct. Translation of the first cistron from both RNAs with either a complete 5'NCR of BVDV or a deletion form is equally efficient in rabbit reticulocyte lysates. However, the complete 5'NCR was able to direct translation of the second cistron in the dicistronic RNA efficiently, whereas the deleted 5'NCR was unable to do so. These preliminary results demonstrate that pestivirus RNA genome is translated by an IRES-mediated mechanism.

5 Conclusions and Future Prospects

Human HCV, a recently added member of Flaviviridae, provides yet another example of a RNA virus which conducts translation by internal ribosome entry. Other flaviviruses except pestiviruses do not share this important translational regulatory scheme. While nothing is known about the mode of viral entry, the early events of infection, and the biological processes that lead to viral hepatitis, the recent findings presented here on IRES-mediated translation hold great promise. These studies are likely to open up avenues of investigation directed toward elucidating fundamental questions relevant to HCV gene expression, the consequences of which manifest in liver disease.

HCV infects hepatocytes, which indicates that the liver cell translational machinery can support IRES-mediated internal initiation of translation. Since these functions rely heavily on the interactions of *trans*-acting cellular factors, future work will focus on mapping structural elements which can function as target sites for RNA-protein interactions. Of interest in this respect are the HCV subtypes which display variable efficiencies of translation (TSUKIYAM-KOHARA et al. 1992). The nucleotide variations of the 5'NCR of these HCV subtypes may provide insight into functionally important structures of internal initiation. Although, we noted that 5'NCR translates with similar efficiencies in both hepatocytic (HepG2, Huh 7) and nonhepatocytic (HeLa) cells, the primary hepatocytes in liver tissues may behave differently. Therefore, it is of considerable importance to explore possible liver cell-specific translation in order to understand the role and mechanism of internal initiation of translation in the liver cell environment.

Finally, the future therapeutic strategies to combat viral hepatitis and perhaps hepatocellular carcinoma may be influenced by consideration of the novel opportunities this translational scheme may offer.

Acknowledgments. The work described in the authors' laboratory is supported by grants from the National Institutes of Health and the Lucille P Markey Charitable trust.

References

Agol VI (1991) The 5'-untranslated region of picornaviral genomes. Adv Virus Res 40: 103–180

Brierley IN, Digard P, Inglis SC (1989) Characterization of an efficient coronavirus ribosomal frameshifting signal: requirement for an RNA pseudoknot. Cell 57: 537–547

Brierley IN, Rolley J, Jenner JA, Inglis SC (1991) Mutational analysis of the RNA pseudoknot component of a coronavirus ribosomal frameshifting signal. J Mol Biol 220: 889–902

Brock KV, Deng R, Riblet SM (1992) Nucleotide sequencing of 5' and 3' termini of bovine viral diarrhea virus by RNA ligation and PCR. J Virol Methods 38: 39–46

Brown EA, Day SP, Jansen RW, Lemon SM (1991) The 5' nontranslated region of hepatitis A virus RNA: secondary structure and elements required for translation in vitro. J Virol 65; 5825–5838

Brown EA, Zhang H, Ping L-H, Lemon SM (1992) Secondary structure of the 5' nontranslated regions of hepatitis C virus and pestivirus genomic RNAs. Nucleic Acids Res 20: 5041–5045

Brown EA, Zajac AJ, Lemon SM (1994) In vitro characterization of an internal ribosomal entry site (IRES) present within the 5' nontranslated region of hepatitis A virus: comparison with the IRES of encephalomyocarditis virus. J Virol 68: 1066–1074

Bukh J, Purcell RH, Miller RH (1992) Sequence analysis of the 5' noncoding region of hepatitis C virus. Proc Natl Acad Sci USA 89: 4942–4946

Chambers TJ, Hahn CS, Galler R, Rice CM (1990) Flavivirus genome organization, expression, and replication. Annu Rev Microbiol 44: 649–688

Chamorro M, Parkin N, Varmus HE (1992) An RNA pseudoknot and an optimal heptameric shift site are required for high efficient ribosomal frameshifting on a retroviral messenger RNA. Proc Natl Acad Sci USA 89: 713–717

Chen P-J, Lin M-H, Tai K-F, Liu P-C, Li C-J, Chen K-S (1992) The Taiwanese hepatitis C virus genome: sequence determination and mapping the 5' termini of viral genomic and antigenomic RNA. Virology 188: 102–113

Choo Q-L, Kuo G, Weiner AJ, Overby LR, Bradley DW, Houghton M (1989) Isolation of a cDNA fragment from a blood-borne non-A, non-B viral hepatitis agent. Science 244: 359–362

Choo Q-L, Richman KH, Han JH, Berger K, Lee C, Dong C, Gallegos C, Coit D, Medina-Selby A, Barr PJ, Weiner AJ, Bradley DW, Kuo G, Houghton M (1991) Genetic organization and diversity of the hepatitis C virus. Proc Natl Acad Sci USA 88: 2451-2455

Collett MS, Larson R, Gold C, Strick D, Anderson DK, Purchio AF (1988) Molecular cloning and nucleotide sequence of the pestivirus bovine viral diarrhea virus. Virology 165: 191–199

Collett MS, Anderson DK, Retzel E (1989a) Comparison of the pestivirus bovine viral diarrhea virus with members of flaviviridae. J Gen Virol 69: 2637–2643

Collett MS, Moenning V, Horzinek M (1989b) Recent advances in pestivirus research. J Gen Virol 70: 253–266

Deng R, Brock KV (1993) 5' and 3'untranslated regions of pestivirus genome: primary and secondary structure analyses. Nucleic Acids Res 21: 1949–1957

Duke GM, Hoffman MA, Palmenberg AC (1992) Sequence and structural elements that contribute to efficient encephalomyocarditis virus RNA translation. J Virol 66: 1602–1609

Feng YX, Yuan H, Rein A, Levin JG (1992) Bipartite signal for read-through suppression in murine leukemia virus mRNA: an eight-nucleotide purine-rich sequence immediately downstream of the gag termination codon followed by an RNA pseudoknot. J Virol 66: 5127–5132

Glass MJ, Summers DF (1993) Identification of a trans-acting activity from liver that stimulates hepatitis A virus translation in vitro. Virology 193: 1047–1050

Glass MJ, Jia X-Y, Summers DF (1993) Identification of the hepatitis A virus internal ribosome entry site: in vivo and in vitro analysis of bicistronic RNAs containing the HAV 5' noncoding region. Virology 193: 842–852

Grakoui A, Wychowski C, Lin C, Feinstone SM, Rice CM (1993) Expression and identification of hepatitis C virus polyprotein cleavage products J Virol 67: 1385–1395

Han JH, Houghton M (1992) Group specific sequences and conserved secondary structures at the 3' end of HCV genome and its implication for viral replication. Nucleic Acids Res 20: 3520

Han JH, Shyamala V, Richman KH, Brauer MJ, Irvine B, Urdea MS, Tekamp-Olson P, Kuo G, Choo Q-L, Houghton M (1991) Characterization of the terminal regions of hepatitis C viral RNA: identification of conserved sequences in the 5' untranslated region and poly(A) tails at the 3' end. Proc Natl Acad USA 88: 1711–1715

Hijikata M, Mizushima H, Tanji Y, Komoda Y, Hirowatari T, Akagi T, Kato N, Kimura K, Shimotohno K (1993) Proteolytic processing and membrane association of putative nonstructural proteins of hepatitis C virus. Proc Natl Acad Sci USA 90: 10773–10777

Houghton M, Weiner A, Han J, Kuo G, Choo Q-L (1991) Molecular biology of the hepatitis C viruses: implications for diagnosis, development and control of viral disease. Hepatology 14: 381–388

Iizuka N, Najita L, Franzusoff A, Sarnow P (1994) Cap-dependent and cap-independent translation of mRNA in cell-free extracts prepared from Saccharomyces cerevisiae. Mol Cell Biol (in press)

Inchauspe G, Zebedee S, Lee D-H, Sugitani M, Nasoff M, Prince AM (1991) Genomic structure of the human prototype strain H of hepatitis C virus: comparison with American and Japanese isolates. Proc Natl Acad Sci USA 88: 10292–10296

Jackson RJ (1991) Potassium salts influence the fidelity of mRNA translation initiation in rabbit reticulocyte lysates: unique features of encephalomyocarditis virus RNA translation. Biochim Biophys Acta 1088: 345–358

Jackson RJ, Howell MT, Kaminiski A (1990) The novel mechanism of initiation of picornavirus RNA translation. Trends Biochem Sci 15: 477–483

Jang SK, Wimmer E (1990) Cap-independent translation of encephalomyocarditis virus RNA: structural elements of the internal ribosomal entry site and involvement of a cellular 57-KD RNA-binding protein. Genes Dev 4: 1560–1572

Jang SK, Davies MV, Kaufman RJ, Wimmer E (1989) Initiation of protein synthesis by internal entry of ribosomes into the 5' nontranslated region of encephalomyocarditis virus RNA in vivo. J Virol 63: 1651-1660

Jia X-Y, Scheper G, Brown D, Updike W, Harmon S, Richards D, Summers D, Erenfeld E (1991) Translation of hepatitis A virus RNA in vitro: aberrant internal initiations influenced by 5' noncoding region. Virology 182: 712–722

Kaminski A, Howell MT, Jackson RJ (1990) Initiation of encephalomyocarditis virus RNA translation: the authentic initiation site is not selected by a scanning mechanism. EMBO J 9: 3753–3759

Kaminski A, Belsham GJ, Jackson RJ (1994) Translation of encephalomyocarditis virus RNA: parameters influencing the selection of the internal initiation site. EMBO J 13: 1673–1681

Kato N, Hijikata M, Ootsuyama Y, Nakagawa M, Ohkoshi S, Sugimura T, Shimotohno K (1990) Molecular cloning of the human hepatitis C virus genome from Japanese patients with non-A, non-B hepatitis. Proc Natl Acad Sci USA 87: 9524–9528

Kettinen H, Grace K, Grunert S, Clarke B, Rowlands D, Jackson RJ (1993) Mapping of the internal ribosome entry site at the 5' end of the hepatitis C virus genome. Proceedings of the International Symposium on Viral Hepatitis and Liver Disease. Tokyo (In press)

Koonin EV (1993) Computer-assisted identification of a putative methyltransferase domain in NS5 protein of flaviviruses and 12 protein of reovirus. J Gen Virol 74: 733–740

Kozak M (1992) Regulation of translation in eukaryotic system. Annu Rev Cell Biol 8: 197–225

Kuhn R, Luz N, Beck E (1990) Functional analysis of the internal translation initiation site of foot-and-mouth disease virus. J Virol 64: 4625–4631

Le S-Y, Chen J-H, Sonenberg N, Maizel JV (1992) Conserved tertiary structure elements in the 5' untranslated region of human enteroviruses and rhinoviruses. Virology 191: 858–866

Le S-Y, Chen J-H, Sonenberg N, Maizel JV Jr (1993) Conserved tertiary structural elements in the 5' nontranslated region of cardiovirus, aphthovirus and hepatitis A virus RNAs. Nucleic Acids Res 21: 2445–2451

Le S-Y, Sonenberg N, Maizel JV Jr (1994) Distinct structural elements and internal entry of ribosomes in mRNA3 encoded by infectious bronchitis virus. Virology 198: 405–411

Lee KAW, Sonenberg N (1982) Inactivation of cap-binding proteins accompanies the shut-off of host protein synthesis by poliovirus. Proc Natl Acad Sci USA 79: 3447–3451

Leathers V, Tanguay R, Kobayashi M, Gallie DR (1993) A phylogenetically conserved sequence within viral 3' untranslated RNA pseudoknots regulates translation. Mol Cell Biol 13: 5331–5347

Liu DX, Inglis SC (1992) Internal entry of ribosomes on a tricistronic mRNA encoded by infectious bronchitis virus. J virol 66: 6143–6154

Meerovitch K, Nicholson R, Sonenberg N (1991) In vitro mutational analysis of cis-acting RNA translational elements within the poliovirus type 2 5' untranslated region. J Virol 65: 5895–5901

Miller RH, Purcell RH (1990) Hepatitis C virus shares amino acid sequence similarity with pestiviruses and flaviviruses as well as members of two plant virus supergroups. Proc Natl Acad Sci USA 87: 2057–2061

Nicholson R, Pelletier J, Le S-Y, Sonenberg N (1991) Structural and functional analysis of the ribosome landing pad of poliovirus type 2: in vivo translational studies. J Virol 65: 5886–5894

Okamoto H, Okada S, Sugiyama Y, Kurai K, Lizuka H, Machida A, Miyakawa Y, Mayumi M (1991) Nucleotide sequence of the genomic RNA of hepatitis C virus isolated from a human carrier: comparison with reported isolates for conserved and divergent regions. J Gen Virol 72: 2697–2704

Pelletier J, Sonenberg N (1988) Internal initiation of translation of eukaryotic mRNA directed by a sequence derived from poliovirus RNA. Nature 334: 320–325

Percy N, Belsham GJ, Brangwyn JK, Sullivan M, Stone DM, Almond JW (1992) Intracellular modifications induced by poliovirus reduce the requirement for structural motifs in the 5' noncoding region of the genome involved in internal initiation of protein synthesis. J Virol 66: 1695–1701
Pestova TV, Hellen CUT, Wimmer E (1991) Translation of poliovirus RNA: role of an essential cis-acting oligopyrimidine element within the 5' nontranslated region and involvement of a cellular 57-kilodalton protein. J Virol 65: 6194–6204
Pilipenko EV, Blinov VM, Ramanova LI, Sinyakov AN, Maslova SV, Agol VI (1989) Conserved structural domains in the 5'-untranslated region of picornaviral genomes: an analysis of the segment controlling translation and neurovirulence. Virology 168: 201–209
Pilipenko EV, Gmyl AP, Maslova SV, Svitkin YV, Sinyakov AN, Agol VI (1992) Prokaryoticlike cis elements in the cap-independent internal initiation of translation on picornavirus RNA. Cell 68: 119–131
Plagemann PGW (1991) Hepatitis C virus. Arch Virol 120: 165–180
Pletnev AG, Yamshchikov VF, Blinov VM (1990) Nucleotide sequence of the genome and complete amino acid sequence of the polyprotein of tick-borne encephalitis virus. Virology 174: 250–263
Poole TL, Wang C, Popp RA, Potgieter LND, Siddiqui A, Collett MS (1995) Pestivirus translation initiation by internal ribosome entry. Virology 206: 750–754
Reynolds JE, Grace K, Clarke BE, Rowlands DJ, Kaminski A, Jackson R (1995) Unusual features of internal initiation of translation of hepatitis C virus RNA (submitted)
Rice CM, Strauss EG, Strauss HJ (1986) Structure of the flavivirus genome. In: Schlessinger S, Schlessinger MJ (eds) The togaviridae and flaviviridae. Plenum, New York, pp 279–326
Ruiz J, Sangro B, Cuende JI, Beloqui O, Riezu-Boj JI, Herrero JI, Prieto J (1992) Hepatitis B and C viral infections in patients with hepatocellular carcinoma. Hepatology 16: 637–641
Saito I, Miyamura T, Ohbayashi A, Harada H, Katayama T, Kikuchi S, Watanabe S, Koi TY, Onji M, Ohta Y, Choo Q-L, Houghton M, Kuo G (1990) Hepatitis C virus infection is associated with the development of hepatocellular carcinoma. Proc Natl Acad Sci USA 87: 6547–6549
Schimmel P (1989) RNA pseudoknots that interact with components of the translation apparatus. Cell 58: 9–12
Selby MJ, Choo Q-L, Berger K, Kuo G, Glazer E, Eckart M, Lee C, Chien D, Kuo C, Houghton M (1993) Expression, identification and subcellular localization of the proteins encoded by the hepatitis C viral genome. J Gen Virol 74: 1103–1113
Shamoo Y, Tarn A, Konigsberg WH, Williams KR (1993) Translational repression by the bacteriophage T4 gene 32 protein involves specific recognition of an RNA pseudoknot structure. J Mol Biol 232: 89–104
Shih DS, Park I-W, Evans CL, Jaynes JM, Palmenberg AC (1987) Effects of cDNA hybridization on translation of encephalomyocarditis virus RNA. J Virol 61: 2033–2037
Shimizu YK, Purcell RH, Yoshikura H (1993) Correlation between the infectivity of hepatitis C virus in vivo and its infectivity in vitro. Proc Natl Acad Sci USA 90: 6037–6041
Skinner MA, Racaniello VR, Dunn G, Cooper J, Minor PD, Almond JW (1989) New model for the secondary structure of the 5' non-coding RNA of poliovirus is supported by biochemical and genetic data that also show that RNA secondary structure is important in neurovirulence. J Mol Biol 207: 379–392
Somogyi P, Jenner AJ, Brierley I, Inglis SC (1993) Ribosomal pausing during translation of an RNA pseudoknot. Mol Cell Biol 13: 6931–6940
Takamizawa A, Mori C, Fuke I, Manabe S, Murakami S, Fujita J, Onishi E, Andoh T, Yoshida I, Okayama H (1991) Structure and organization of the Hepatitis C virus genome isolated from human carriers. J Virol 65: 1105–1113
Tanaka T, Kato N, Nakagawa M, Ootruyama Y, Cho MJ, Nakazawa T, Hijikata M, Ishimura Y, Shimotohno K (1992) Molecular cloning of hepatitis C virus genome from a single Japanese carrier: sequence variation within the same individual and among infected individuals. Virus Res 23: 39–53
Tsukiyama-Kohara K, Iizuka N, Kohara M, Nomoto A (1992) Internal ribosome entry site within hepatitis C virus RNA. J Virol 66. 1476–1483
Tu C, Tzeng T-H, Bruenn JH (1992) Ribosomal movement impeded at a pseudoknot required for frameshifting. Proc Natl Acad Sci USA 89: 8636–8640
Tzeng T-H, Tu C-L, Bruenn JA (1992) Ribosomal frameshifting require a pseudoknot in the Saccharomyces cerevisiae double-stranded RNA virus J Virol 66: 999-1006
Wang C, Le SY, Siddiqui A (1995) Functional role of an RNA pseudoknot structure in internal initiation of translation of hepatitis C virus RNA genome. (submitted for publication)

Wang C, Sarnow P, Siddiqui A (1993) Translation of human hepatitis C virus RNA in cultured cells in mediated by an internal ribosome-binding mechanism. J Virol 67: 3338–3344
Wang C, Sarnow P, Siddiqui A (1994) A conserved helical element is essential for the internal initiation of translation of the hepatitis C virus RNA. J Virol 68: 7301–7307
Wills NM, Gesteland RF, Atkins JF (1991) Evidence that a downstream pseudoknot is required for translational read-through of the moloney murine leukemia virus gag stop codon. Proc Natl Acad Sci USA 88: 6991–6995
Wimmer E, Hellen CUT, Cao X (1993) Genetics of poliovirus. Annu Rev Genet 27: 353–436
Wiskerchen M, Belzer SK, Collett MS (1991) Pestivirus gene expression: the first protein product of the bovine viral diarrhea virus large open reading frame, p20, possesses proteolytic activity. J Virol 65: 4509–4514
Yoo BJ, Spaete RR, Geballe AP, Selby M, Houghton M, Han JH (1992) 5' end-dependent translation initiation of Hepatitis C viral RNA and the presence of putative positive and negative translational control elements within the 5' untranslated region. Virology 191: 889–899

Cap-Independent Translation in Adenovirus Infected Cells

R.J. SCHNEIDER

1 Adenovirus Life Cycle

Adenoviruses (Ads) are DNA viruses that infect humans, animals and birds, with different serotypes displaying different tissue tropisms (BELADI 1972). Ad was originally isolated because infection results in cytopathic effects and alterations in basic cellular metabolism. The Ad genome is temporally organized into early and late transcription units that are activated before or with the onset of viral DNA replication, respectively. Six early transcription units encode products required for productive viral replication and transformation of the infected cell. Regions E1A and E1B are required for cellular transformation and transactivation of the other viral transcription units (FLINT and SHENK 1989). Regions E2A and E2B are required for adenoviral DNA replication. Regions E3 and E4 are required for a variety of early viral functions, including suppression of histocompatability antigen expression (reviewed in WOLD and GOODING 1991), transcriptional transactivation and regulation of nuclear to cytoplasmic transport of cellular and viral mRNAs

Department of Biochemistry and Kaplan Cancer Center, New York University Medical School, New York, NY 10016, USA (reviewed in Schneider and Zhang 1993). The products of the early transcription units comprise only a very minor proportion of cellular mRNA and protein synthesis, and there is no evidence for selective viral translation or inhibition of cell protein synthesis during the early part of the Ad life cycle.

The late phase of Ad infection is marked by viral DNA replication, initiating at 10–16 h after infection. Whereas early viral transcription is initiated from promoters dispersed throughout the viral genome, there is a single major late promoter (MLP), located at 16.4 map units on the viral genome, that is activated by DNA replication. The MLP generates five families of late transcripts (L1–L5) by differential splicing and polyadenylation of a large primary transcript that terminates within the right end of the genome at 99 map units (reviewed in Ginsberg 1984). Every late viral mRNA contains a common 5' noncoding region of 200 nucleotides called the tripartite leader (Berget et al. 1977), derived by splicing three small exons located upstream of the late transcripts. Most of the late Ad mRNAs encode structural polypeptides involved in packaging viral genomic DNAs that comprise the viral particle. Ad also synthesizes large amounts of two viral encoded RNA polymerase III products during late infection called virion-associated (VA) RNAs I and II (reviewed in Thimmapaya et al. 1993; Mathews and Shenk 1991). VA RNA I is required for translation of mRNAs at late times during infection because it counters a cellular antiviral response mediated by the interferon stimulated p68 kinase.

Ad infection of cells in culture occupies a life cycle lasting about 2–4 days, during which time large quantities of late viral polypeptides and infectious particles are produced. The late phase of infection is associated with almost exclusive translation of late Ad mRNAs and inhibition of cell protein synthesis (reviewed in Schneider and Zhang 1993), and impaired transport of cellular mRNAs from the nucleus to cytoplasm (Beltz and Flint 1979). Cellular synthesis of DNA, RNA and proteins is usurped for the production and assembly of viral particles which are released when cell lysis occurs. The block to cellular protein synthesis occurs during progression into the late phase of Ad replication. Late viral mRNAs generally represent the majority (~90%–95%) found in polyribosomes, but only a fraction of the total cytoplasmic pool of mRNAs (reviewed in Schneider and Zhang 1993). Therefore cellular mRNAs are suppressed from translating and late Ad tripartite leader mRNAs are preferentially used.

Ad inhibition of cellular translation is not related to the viral block in transport of host mRNAs from the nucleus to cytoplasm. Studies showed that the cytoplasmic abundance of most cell mRNAs does not significantly decline during late infection (Babich et al. 1983). It was also shown that Ad inhibition of cellular protein synthesis can be prevented by the drug 2-aminopurine without relieving the normal block in transport of host mRNAs (Huang and Schneider 1990). These results imply that translation of cellular mRNAs is specifically prevented during late Ad infection while late Ad mRNAs are preferentially utilized.

2 Translational Regulation by the Tripartite Leader 5' Noncoding Region

2.1 Adenovirus Tripartite Leader 5' Noncoding Region Is Required for Translation of mRNAs During Late Viral Infection

At late times after infection most late mRNAs are transcribed from the MLP, giving rise to five families of 3' coterminal mRNAs, all of which share a common 5' noncoding region called the tripartite leader. Mutational analysis of the tripartite leader and its reconstruction into recombinant Ad vectors demonstrated that the intact leader was required for translation of mRNAs at late but not early times after Ad infection (LOGAN and SHENK 1984; BERKNER and SHARP 1985). The tripartite leader was also shown to enhance translation of mRNAs in transfected cells (KAUFMAN 1985). These studies led to the suggestion that the tripartite leader may be involved in preferential translation during late Ad infection.

2.2 The Tripartite Leader Reduces the Requirement for Cap Binding Protein Complex (eIF-4F)

Several studies found that late Ad mRNAs possess unusual translation properties in that they are resistant to inhibition by super-infecting poliovirus (CASTRILLO and CARRASCO 1987; DOLPH et al. 1988). Poliovirus infection results in inhibition of cap-dependent cellular protein synthesis which correlates with proteolytic degradation of a 220 kDa polypeptide (p220) (ETCHISON et al. 1982; GRIFO et al. 1983), a component of initiation factor eIF-4F. This factor is a cap-dependent RNA helicase that stimulates protein synthesis by unwinding the 5' end of mRNAs, facilitating binding of 40S ribosomes to mRNA (SONENBERG et al. 1982; RAY et al. 1985; LAWSON et al. 1986; reviewed in FREDERICKSON and SONENBERG 1983). eIF-4F contains three proteins: eIF-4E, a 24 kDa protein which specifically binds cap structures; eIF-4A, a 45 kDa ATP-dependent RNA helicase; p220, a 220 kDa protein of unknown function (reviewed in RHOADS 1988; THACH 1992). Degradation of p220 during poliovirus infection prevents or alters cap-dependent RNA helicase activity associated with eIF-4F, which is not essential for translation of picornaviral mRNAs because they internally bind ribosomes (PELLETIER and SONENBERG 1988; JANG et al. 1988). As might be expected, eIF-4F helps to overcome the barrier to translation conferred by stable secondary structure in 5' noncoding regions. Stable secondary structure in the 5' noncoding region correlates with an increased requirement and decreased binding of eIF-4F, while relaxed 5' structure enhances the binding and decreases the requirement for eIF-4F (RAY et al. 1985; LAWSON et al. 1986; BROWNING et al. 1988; FLETCHER et al. 1990). Therefore, the ability of an mRNA to translate in the absence of eIF-4F activity, or in the presence of minimal amounts of this factor, represents a unique translation strategy available to only a small number of mRNAs. Picornaviral and several cellular mRNAs have been found to possess little if any requirement for eIF-4F because they internally initiate translation (PELLETIER and SONENBERG 1988; JANG et al.1988; SARNOW 1989; MACEJAK and SARNOW 1991). A few mRNAs require little if any eIF-4F because they possess 5' noncoding regions with minimal 5' secondary structure, such as alflafa mosaic virus 4 (AMV 4) mRNA (FLETCHER et al. 1990), and heat shock hsp70 and hsp83 mRNAs (LINDQUIST and PETERSON 1991; ZAPATA et al. 1991).

The tripartite leader was shown to be sufficient to confer translation independent of eIF-4F activity in poliovirus infected cells (DOLPH et al. 1988). Heterologous (non-Ad) mRNAs containing a cDNA copy of the tripartite leader were found to be efficiently translated when expressed from plasmids in transfected cells in the presence of infecting poliovirus or Ad. Thus, "eIF-4F-independent" translation by the tripartite leader does not require Ad gene products, but rather is conferred solely by the 5' noncoding region.

2.3 The Tripartite Leader Does Not Direct Internal Ribosome Entry

The mechanism by which the tripartite leader reduces or eliminates the requirement for eIF-4F was investigated in detail (DOLPH et al. 1990). Several mechanisms can be envisioned for tripartite leader translational activity. (1) Promotion of internal initiation through an internal ribosome entry site (IRES), as described for picornaviral mRNAs (reviewed in JACKSON 1991). (2) Internal initiation utilizing unique nucleotide sequences or secondary structures in a prokaryotic-type interaction between 18S rRNA and the mRNA. (3) Translation initiating at an unstructured 5' end, as shown for hsp70, hsp83 and AMV 4 mRNAs (reviewed in THACH 1992). (4) Cap-dependent but nonlinear ribosome entry through ribosome "jumping" or "shunting".

The tripartite leader was shown to be incapable of directing internal ribosome entry in a manner described for picornavirus mRNAs. When constructed as the second cistron of a dicistronic mRNA and examined by transfection of plasmids into cells, the tripartite leader could not promote internal translation initiation (DOLPH et al. 1990). However, the picornaviral IRES elements readily direct internal translation from dicistronic mRNAs (PELLETIER and SONENBERG 1988; JANG et al. 1988). Infection of transfected cells with polio or Ad virus also did not alter the inability of the tripartite leader to direct internal ribosome entry. It is therefore unlikely that the tripartite leader facilitates internal ribosome binding in a manner analogous to IRES elements. It is also unlikely, although not fully excluded, that the tripartite leader directs mRNA:rRNA interactions akin to the Shine-Dalgarno sequence of prokaryotes. Although the tripartite leader contains three regions complementary to a conserved hairpin structure in the 3' end of 18S rRNA, deletion of these elements had no detectable effect on translation efficiency or eIF-4F independence (DOLPH et al. 1990). It should be noted, though, that the effect of coordinate deletion of all three elements on translation was not examined, leaving open the possibility that any one element may be sufficient. Nevertheless, evidence was obtained indicating that the 5' end of the tripartite leader contains an unstructured conformation (DOLPH et al. 1990). It was therefore concluded that the tripartite leader probably assumes a conformation independent of the mRNA to which it is attached, but one which provides a relaxed 5' end since its structure was unaltered by the body of the mRNA (ZHANG et al. 1989) and because it confers eIF-4F independence to all mRNAs to which it is linked. Consistent with these results, duplication of the 5' end of the tripartite leader (nucleotides 1–33) in the antisense orientation created a stable hairpin structure, reduced translational efficiency and most importantly rendered the mRNA fully dependent on eIF-4F for its translation (DOLPH et al. 1990). Secondary structure analysis of the RNA indicated that the 5' end of the leader may be unstructured (ZHANG et al. 1989). The first 22 nucleotides and most of leader 1 (1–44 nucleotides) was found to be relatively devoid of structure as determined by enzymatic probing using single-strand specific nucleases. Whether the reduced 5' proximal secondary structure and ribosome shunting are involved in tripartite leader mediated translation initiation will be addressed at the end of this review.

2.4 Adenovirus Dephosphorylation of eIF-4E Impairs Activity of eIF-4F During Late Infection

Since the tripartite leader confers the ability to translate in poliovirus infected cells, apparently independent of eIF-4F, the activity of initiation factor eIF-4F was implicated as a target for Ad control. Early experiments showed that eIF-4F is not inactivated by proteolysis of the p220 component in late Ad infected cells as occurs during poliovirus infection (DOLPH et al. 1988). However, the activity of eIF-4F is also regulated by phosphorylation of the eIF-4E (CBP) component. Reduced phosphorylation of eIF-4E correlates with a decreased ability of the factor to associate with 40S ribosomes (JOSHI-BARVE et al. 1990), and with inhibition of translation during heat shock (DUNCAN et al. 1987; LAMPHEAR and PANNIERS 1991; ZAPATA et al. 1991), and mitosis (BONNEAU and SONENBERG 1987). Increased phosphorylation of eIF-4E correlates with enhanced translation after activation of cells by mitogens or serum (MORLEY and TRAUGH 1989; MARINO et al. 1989; KASPAR et al. 1990), and in cells transformed with the *src* oncogene (FREDERICKSON et al. 1991). Overexpression of eIF-4E in cells also results in aberrant and enhanced cell growth (SMITH et al. 1990), including transformation (DEBENEDETTI and RHOADS 1990; LAZARIS-KARATZAS et al. 1990). Again, phosphorylation of eIF-4E is required for this effect. The phosphorylation of eIF-4E does not greatly alter its ability to bind specifically to cap structures (RYCHLIK et al. 1986; HIREMATH et al 1989). Instead, phosphorylation of eIF-4E may be required to promote eIF-4F association with 40S ribosomes or to enhance ribosome association with mRNA mediated by eIF-4F (JOSHI-BARVE et al. 1990).

Investigation of eIF-4E phosphorylation in late Ad infected cells showed that it is a target for Ad regulation (HUANG and SCHNEIDER 1991). By quantitating the level of eIF-4E which could be labeled in vivo with ^{32}P-orthophosphate, the level in eIF-4E was found to be reduced 10–20 fold at late times after Ad infection compared to uninfected cells. The under-phosphorylation of eIF-4E also occurred with kinetics consistent with the shutoff of cellular protein synthesis during late infection. A previous study demonstrated that treatment of cells with the kinase inhibitor 2-aminopurine could block the shutoff of host protein synthesis during the late phase of Ad infection with only minor (three- to fourfold) reductions in viral DNA and late mRNA synthesis (HUANG and SCHNEIDER 1990). Accordingly, 2-aminopurine was found to prevent dephosphorylation of eIF-4E, consistent with a role for its dephosphorylation in the shutoff of host cell protein synthesis (HUANG and SCHNEIDER 1991). By resolving the steady-state population of unlabeled eIF-4E into phosphorylated and unphosphorylated forms using 2-dimensional isoelectric focusing immunoblot analysis, the fraction of phosphorylated eIF-4E was found to be reduced to about 5% that of the total protein in late Ad infected cells, but dephosphorylation of eIF-4E during infection was largely blocked by treatment of cells with 2-aminopurine. It is important to note that dephosphorylation of eIF-4E is not complete in late Ad infected cells, leaving ~5% of the factor in a phosphorylated form that could conceivably participate in formation of low levels of eIF-4F. Studies indicate that mRNAs lacking 5' proximal secondary structure either require less eIF-4F or recruit it more efficiently (reviewed in THACH 1992). Although the study described above implicated dephosphorylation of eIF-4E in Ad shutoff of cell translation, it did not address whether the tripartite leader permits late Ad mRNAs to translate in the absence of eIF-4F activity, or efficiently recruits the low levels of eIF-4F that remain during the late phase of infection.

2.5 Adenovirus Tripartite Leader Requires Low Levels of eIF-4F for Translation

Studies conducted by Voorma and colleagues (THOMAS et al. 1992) were directed toward determining whether the tripartite leader permits authentic cap-independent translation or efficiently competes for the small amounts of active eIF-4F that remain in a late Ad infected cell. In vitro transcribed mRNAs containing the tripartite leader were compared to mRNAs containing the encephalomyocarditis Virus (EMCV) IRES element for dependence on eIF-4F. Activity of eIF-4F was destroyed in reticulocyte in vitro translation extracts by prior incubation with the foot and mouth disease virus (FMDV) L-protease, which fully degrades the eIF-4F p220 subunit. Whereas IRES mediated translation still occurred in the absence of intact eIF-4F, tripartite leader driven translation was inhibited. Importantly, cleavage of p220 by the poliovirus 2A protease is generally not complete in cells or extracts (e.g., ALVEY et al. 1991; DOLPH et al. 1988; THOMAS et al. 1992), although cleavage by the FMDV L-protease does go to completion and occurs at a different site (THOMAS et al. 1992). These results suggest two different interpretations: (1)

the tripartite leader requires low levels of eIF-4F to facilitate translation, or (2) cleavage of p220 by the L-protease destroys eIF-4F activity, but cleavage by the 2A protease either reduces or alters its activity. The distinction is an important one and should be resolved in future studies because even internal initiation of translation on poliovirus and cowpea-mosaic virus mRNAs has been found to be stimulated by eIF-4F (ANTHONY and MERRICK 1991; THOMAS et al. 1991).

3 Mechanism for Cap-Independent Translation of Late Adenoviral mRNAs Mediated by the Tripartite Leader

A model was proposed (HUANG and SCHNEIDER 1991; reviewed in SCHNEIDER and ZHANG 1993) in which the ability of the tripartite leader to confer translation with little requirement for eIF-4F and viral dephosphorylation of eIF-4E could account for selective translation of late Ad mRNAs (Fig. 1). It was suggested that late Ad mRNAs efficiently recruit the small amounts of active or altered eIF-4F that remain in late infected cells because of their lack of 5' proximal secondary structure. In this regard, the tripartite leader may be similar to AMV-4 mRNA, which also efficiently recruits limited amounts of eIF-4F through its affinity for unstructured 5' ends of capped mRNAs (FLETCHER et al. 1990). The tripartite leader would therefore serve as a means to discriminate between late Ad mRNAs that require minimal eIF-4F and most cell mRNAs that typically possess a larger requirement for the factor to enable translation.

The fact that eIF-4F activity is largely impaired in late Ad infected cells by dephosphorylation of eIF-4E indicates that its ability to promote cap-dependent RNA unwinding must be deficient. In addition to facilitating ribosome-mRNA interactions, eIF-4F might also aid in the 40S ribosome subunit search for an initiating AUG since it displays 5'-to-3' unwinding activity for capped mRNAs, but 3'-to-5' unwinding in the absence of caps (ROZEN et al. 1990). This raises the question as to how the tripartite leader actually promotes initiation of translation. A mechanism that relies purely on ribosome-cap association and subsequent scanning by 40S subunits might be expected to require larger amounts of a cap-dependent eIF-4F helicase activity than one utilizing internal ribosome binding. However, one mechanism that potentially requires lower levels of eIF-4F, internal ribosome binding, is not carried out by the tripartite leader in the context of a discistronic mRNA (DOLPH et al. 1990; JANG et al. 1989).

Since even internal translation initiation benefits from eIF-4F activity (ANTHONY and MERRICK 1991), all forms of ribosome initiation may require RNA unwinding regardless of whether they occur in cap-dependent or independent manner. This has been confirmed in a recent report, in which a dominant-negative inhibitor of eIF-4A, the RNA helicase component of eIF-4F (GRIFO et al. 1983; ROZEN et al. 1990), was shown to prevent initiation complex formation by ribosomes on both cap-dependent and independent mRNAs (PAUSE et al. 1994). Thus, translation of

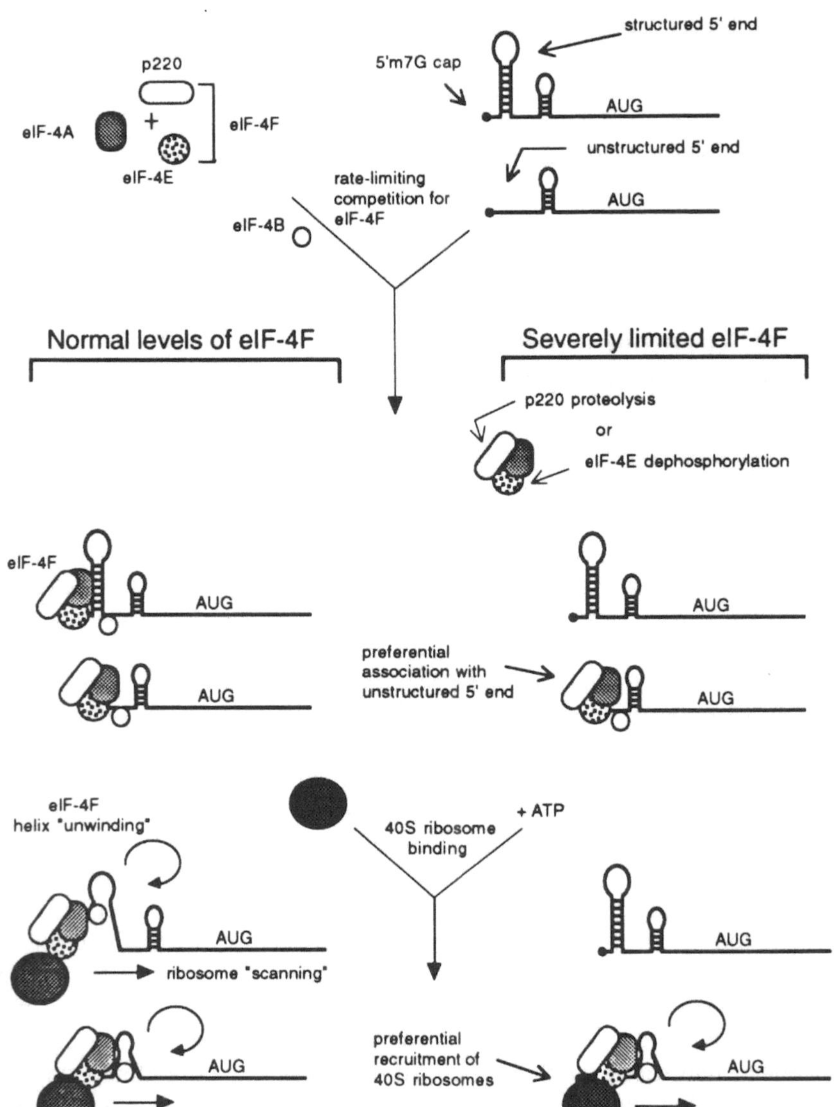

Fig. 1. Model for suppression of cellular protein synthesis and selctive translation of adenovirus (Ad) late mRNAs. Cap recognition by eIF-4F (eIF-4E + eIF-4A + p220) is a rate-limiting step in translation initiation. Dephosphorylation of eIF-4E during late Ad infection severely reduces the pool of active eIF-4F. Tripartite leader mRNAs recruit the small amounts of active eIF-4F very efficiently due to relaxed 5' proximal secondary stucture, effectively bypassing a limiting step in translation of host cell mRNAs. (Figure originally published in ZHANG and SCHNEIDER 1993)

all mRNAs is likely to involve some RNA unwinding activity including those like the tripatite leader or AMV-4 which are less dependent on cap structures. These findings indicate that non-cap-dependent RNA unwinding occurs during translation regardless of whether 40S ribosomes initiate via scanning by binding to cap structures, or by other mechanisms such as direct binding to elements within the 5' noncoding region. They also raise the issue as to whether dephosphorylation of eIF-4F alters rather than inactivates eIF-4F, perhaps making eIF-4F more cap-independent in activity. Consequently, in late Ad infected cells the combined dephosphorylation of eIF-4E and the reduced dependence of the tripartite leader on eIF-4F activity could promote a cap-dependent, but nonscanning, mechanism for ribosome initiation utilizing signals within the 5' noncoding region.

Recent studies in the author's lab have been directed toward detemining whether the tripartite leader utilizes a cap-dependent scanning mechanism to initiate ribosomes or whether other mechanisms may be at play that theoretically are not as demanding of cap-dependent eIF-4F activity. One mechanism, internal ribosome entry, has been excluded for the tripartite leader as described above. However, another semi- or nonlinear mechanism for initiation has also been described, known as ribosome shunting or jumping, in which 40S ribosome subunits require cap-dependent recognition of mRNAs, undergo limited scanning from the 5' end and then undergo nonlinear translocation through the 5' noncoding region to the initiating AUG. Although rare in eukaryotic cells, results consistent with initiation via ribosome shunting have been described for translation of the cauliflower mosaic virus 35S mRNA FUTTERER et al. 1993) and possibly for Sendai Virus X gene mRNA (CURRAN and KOLAKOFSKY 1988), as well as for several prokaryotic mRNAs (BENHAR and ENGELBERG-KULKA 1993; HUANG et al. 1988). A hallmark of ribosome shunting is a dependence on cap-initiated scanning which can be blocked by introduction of 5' proximal secondary structure. Translation is unimpaired if secondary structure is inserted downstream in a region that is bypassed by ribosomes.

To determine whether the tripartite leader might promote ribosome shunting, secondary structure was inserted in two locations downstream of the cap site, at 30 and 200 nucleotides (the 3' end) of the tripartite leader. As shown in Fig. 2, stem-loop structures were created by ligating seven copies of a 12 nucleotide *Bam*HI linker in tandem ($\Delta G > -80$ kcal/mol). A control 5' noncoding region was constructed in which an initiating AUG was placed in the unstructured 5' end of the leader followed by the stem-loop structure. Cells were transfected with plasmids encoding the wild-type and variant tripartite leaders coupled to the hepatitis B virus surface antigen (HBsAg) coding region as a marker. Levels of mRNA were determined by northern analysis, and levels of HBsAg protein by immunoprecipitation of ^{35}S-methionine-labeled protein extracts and SDS-polyacrylamide gel electrophoresis. All constructs synthesized equivalent amounts of mRNAs, thereby excluding effects of mRNA accumulation. However, when levels of HBsAg were examined, it was clear that secondary structure did not block translation when inserted downstream at the 3' end of the tripartite leader. Moreover, introduction of an initiation codon in the 5' region of the tripartite leader eliminated the ability of 40S ribosomes to translate downstream of the stem-loop structure, probably because it promoted formation of 80S ribosomes which were tethered to the mRNA. These results are therefore consistent with, but do not

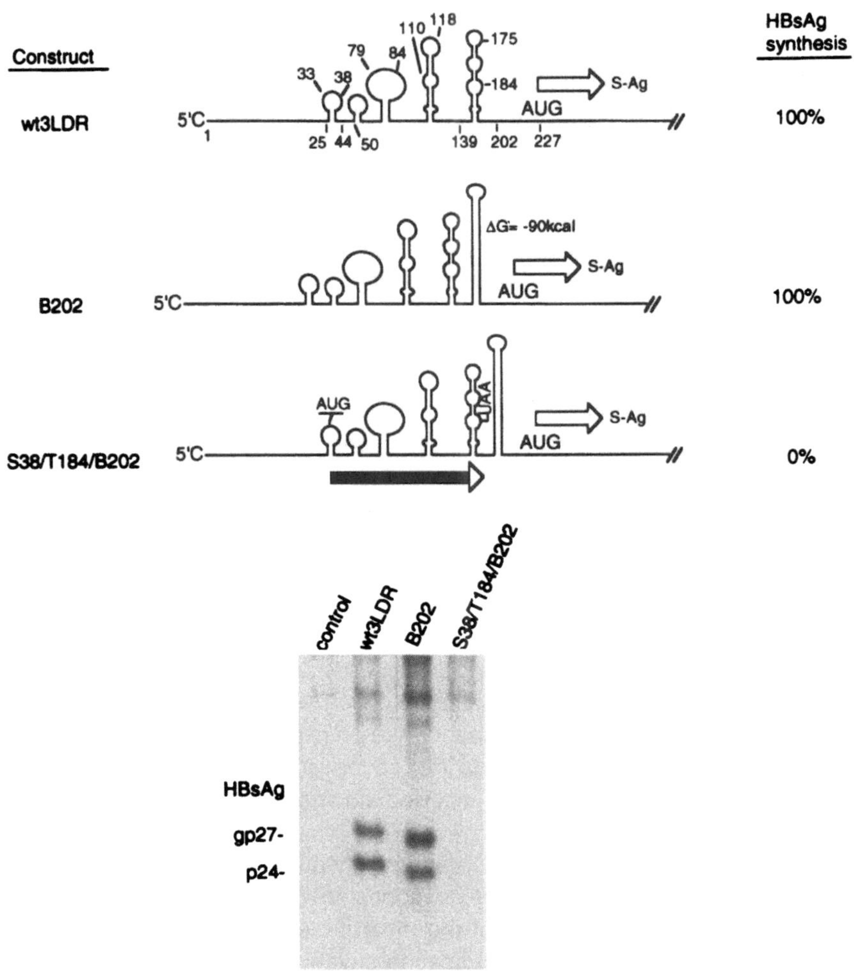

Fig. 2. Evidence for ribosome shunting in the adenovirus (Ad) tripartite leader. Seven tandem copies of 12 nucleotide *Bam*HI linkers or an initiating AUG were inserted in the tripartite leader as shown. mRNAs were expressed from plasmids under the control of the Ad major late promoter (MLP) and E1B splice/polyadenylation signals. Immunoprecipitation analysis of ^{35}S-methionine labeled HBsAg protein was performed as described previously. (From DOLPH et al. 1988)

prove, that the tripartite leader facilitates the cap-dependent but nonlinear mechanism for ribosome initiation known as ribosome shunting.

In principle, ribosome shunting might facilitate translation because it possibly confers a reduced requirement for translation initiation factors eIF-4F or eIF-4A, although there is not yet any strong experimental evidence regarding this point. Additionally, ribosome shunting may aid in distinguishing late Ad mRNAs from cellular mRNAs after dephosphorylation of eIF-4E, thereby enhancing their selective translation. Ongoing studies are now addressing the molecular mechanism for ribosome shunting and its function in translational control by the Ad tripartite leader.

Acknowledgments. The author's work described in this review was supported by a grant from the National Institutes of Health (CA-42357).

References

Alvey JC, Wyckoff EE, Yu SF, Lloyd R, Ehrenfeld E (1991) Cis and trans cleavage activities of poliovirus 2A protease expressed in Escherichia coli. J Virol 65: 6077–6083

Anthony DD, Merrick WC (1991) Eucaryotic initiation factor eIF-4F. J Biol Chem 266: 10218–10226

Babich A, Feldman CT, Nevins JR, Darnell JE, Weinberger C (1983) Effect of adenovirus on metabolism of specific host mRNAs: transport control and specific translational discrimination. Mol Cell Biol 3: 1212–1221

Beladi I (1972) Strains of human viruses. In: Majer M, Plotkin SA (eds) Adenoviruses. Karger, Basel

Beltz GA, Flint SJ (1979) Inhibition of Hela cell protein synthesis during adenovirus infection. J Mol Biol 131: 353–373

Benhar I, Engelberg-Kulka H (1993) Frameshifting in the expression of the E. coli trpR gene occurs by the bypassing of a segment of ts coding region. Cell 72: 121–130

Berget SM, Moore C, Sharp P (1977) Spliced segments at the 5' terminus of Ad2 late mRNA. Proc Natl Acad Sci USA 74: 3171–3175

Berkener KE, Sharp PA (1985) Effect of tripartite leader on synthesis of a nonviral protein in adenovirus 5' recombinant. Nucleic Acids Res 13: 841–857

Bonneau AM, Sonenberg N (1987) Involvement of the 24kd cap-binding protein in regulation of protein synthesis in mitosis. J Biol Chem 262: 11134–11139

Browning KS, Fletcher L, Ravel JM (1988) Evidence that the requirements for ATP and wheat germ initiation factors 4A and 4F are affected by a region of satellite tobacco necrosis virus RNA that is 3' to the ribosomal binding site. J Biol Chem 262: 8380–8383

Castrillo JL, Carrasco L (1987) Adenovirus late protein synthesis is resistant to the inhibition of translation induced by poliovirus. J Biol Chem 262: 7328–7334

Curran J, Kolakofsky D (1988) Scanning independent ribosomal initiation of the Sendai virus X protein. EMBO J 7: 2869–2874

DeBenedetti A, Rhoads RE (1990) Overexpression of eukaryotic protein synthesis initiation factor 4E in Hela cells results in aberrant growth and morphology. Proc Natl Acad Sci USA 87: 8212–8216

Dolph PJ, Racaniello V, Villamarin A, Palladino F, Schneider RJ (1988) The adenovirus tripartite leader eliminates the requirement for cap binding protein during translation initation. J Virol 62: 2059–2066

Dolph PJ, Huang J, Schneider RJ (1990) Translation by the adenovirus tripartite leader: Elements which determine independence from cap-binding protein complex. J Virol 64: 2669–2677

Duncan R, Milburn SC, Hershey JWB (1987) Regulated phosphorylation and low abundance of Hela cell initiation factor eIF-4F suggest a role in translational control. J Biol Chem 262: 380–388

Etchison D, Milburn SC, Edery I, Sonenberg N, Hershey JWB (1982) Inhibition of Hela cell protein synthesis following poliovirus infection correlates with the proteolysis of a 220,000 dalton polypeptide associated with eucaryotic initiation factor 3 and a cap binding protein complex. J Biol Chem 257: 14806–14810

Fletcher L, Corbin SD, Browning KG, Ravel JM (1990) The absence of a m^7G cap on Beta-globin mRNA and alfalfa mosaic virus 4 increases the amounts of initiation factor 4F required for translation. J Biol Chem 265: 19582–19587

Flint J, Shenk T (1989) Adenovirus E1a protein: paradigm viral transactivator. Annu Rev Genet 23: 141–161

Frederickson RM, Sonenberg N (1993) eIF-4E phosphorylation and the regulation of protein synthesis. In: Ilan J (ed) Translational regulation of gene expression, vol2. Plenum, New York

Frederickson RM, Montine KS, Sonenberg N (1991) Phosphorylation of eucaryotic translation initiation factor 4E is increased in Src-transformed cell lines. Mol Cell Biol 11: 2896–2900

Futterer J, Kiss-Laszlo Z, Hohn T (1993) Nonlinear ribosome migration on cauliflower mosaic virus 35S RNA. Cell 73: 789–802

Ginsberg HS (1984) The adenoviruses. Plenum, New York

Grifo JA, Tahara SM, Morgan MA, Shatkin AJ, Merrick WC (1983) New initiator activity required for globin mRNA translation. J Biol Chem 258: 5804–5810

Hiremath LS, Hiremath ST, Rychlik W, Joshi S, Domier LL, Rhoads RE (1989) In vitro synthesis, phosphorylation and localization on 48S initiation complexes of human protein synthesis initiation factor 4E. J Biol Chem 264: 1132–1138

Huang J, Schneider RJ (1990) Adenovirus inhibition of cellular protein synthesis is prevented by the drug 2-aminopurine. Proc Natl Acad Sci USA 87: 7115–7119

Huang J, Schneider RJ (1991) Adenovirus inhibition of cellular protein synthesis involves inactivation of cap binding protein. Cell 65: 271–280

Huang WM, Ao SZ, Casjens S, Orlandi R, Zeikus R, Weiss R, Winge D, Fang M (1988) A persistent untranslated sequence within bacteriophage T4 DNA topoisomerase gene 60. Science 239: 1005–1012

Jackson RJ (1991) Initiation without an end. Nature 353. 14–15

Jang SK, Kraussilch HG, Nicklin MJH, Duke GM, Palmenberg AC, wimmer E (1988) A segment of the 5' nontranslated region of encephalomyocarditis virus RNA directs internal entry of ribosomes during in vitro translation. J Virol 62: 2636–2643

Jang SK, Davies MV, Kaufman RJ, Wimmer E (1989) Initiation of protein synthesis by internal entry of ribosomes into the 5' nontranslated region of encephalomyocarditis virus RNA in vivo. J Virol 63: 1651–1660

Joshi-Barve S, Rychlik W, Rhoads RE (1990) Alteration of the major phosphorylation site of eukaryotic protein synthesis initiation factor 4E prevents its association with the 48S initation complex. J Biol Chem 265: 2979–2983

Kaspar R, Rychlik W, White MW, Rhoads RE, Morris DR (1990) Simultaneous cytoplasmic redistribution of ribosomal protein L32 mRNA and phosphorylation of eukaryotic initiation factor 4E after mitogenic stimulation of Swiss 3T3 cells. J Biol Chem 265: 3619–3622

Kaufman RJ (1985) Identification of the components necessary for adenovirus translational control and their utilization in cDNA expression vectors. Proc Natl Acad Sci USA 82: 689–693

Lamphear BJ, Panniers R (1991) Heat shock impairs the interaction of cap binding protein complex with 5' mRNA cap. J Biol Chem 266: 2789–2794

Lawson TG, Ray BK, Dodds JT, Grifo JA, Abramson RD, Merrick WC, Betsch DF, Weith HL, Thach RE (1986) Influence of 5' proximal secondary structure on the translational efficiency of eukaryotic mRNAs and on their interaction with initiation factors. J Biol Chem 261: 13979–13989

Lazaris-Karatzas A, Montine KS, Sonenberg N (1990) Malignant transformation by a eukaryotic initiation factor subunit that binds to mRNA 5' cap. Nature 345: 544–547

Lindquist S, Peterson R (1991) Selective translation and degradtion of heat shock messenger RNAs in drosophila. Enzyme 44: 147–166

Logan J, Shenk T (1984) Adenovirus tripartite leader sequence enhances translation of mRNAs late after infection. Proc Natl Acad Sci USA 81: 3655–3659

Macejak DG, Sarnow P (1991) Internal initiation of translation mediated by the 5' leader of a cellular mRNA. Nature 353. 90–94

Marino MW, Pfeffer LM, Guidon PT, Donner DB (1989) Tumor necrosis factor induces phosphorylation of a 28kd mRNA cap-binding protein in human cervical carcinoma cells. Proc Natl Acad Sci USA 86: 8417–8421

Mathews MB, Shenk T (1991) Adenovirus virus-associated RNA and translational control. J Virol 65: 5657–5662

Morley SJ, Traugh JA (1989) Phorbol esters stimulate phosphorylation of eukaryotic initiation factors 3, 4B and 4F. J Biol Chem 264: 2401–2404

Pause A, Mehtot N, Svitkin Y, Merrick WC, Sonenberg N (1994) Dominant negative mutants of mammalian initation factor eIF-4A define a critical role for eIF-4F in cap-dependent and cap-independent initation of translation. EMBO J 13: 1205–1215

Pelletier J, Sonenberg n (1988) Internal initiation of translation of eukaryotic mRNA directed by a sequence derived from poliovirus RNA. Nature 334: 320–325

Ray BK, Lawson TG, Kramer JC, Cladarns MH, Grifo JA, Abramson RD, Merrick WC, Thach RE (1985) ATP dependent unwinding of messenger RNA structure by eukaryotic initiation factors. J Biol Chem 260: 7651–7658

Rhoads RE (1988) Cap recognition and the entry of mRNA into the protein synthesis initiation cycle. Trends Biochem Sci 13: 52–56

Rozen F, Edery I, Meerovitch K, Dever TE, Merrick WC, Sonenberg N (1990) Bidirectional RNA helicase activity of eukaryotic translation initiation factor 4A and 4F. Mol Cell Biol 10: 1134–1144

Rychlik W, Gardner PR, Vanaman TC, Rhoads RE (1986) Structural analysis of the messenger RNA cap-binding protein. J Biol Chem 261: 71–75

Sarnow P (1989) Translation of glucose regulated protein 78/immunoglobulin heavy chain binding protein mRNA is increased in poliovirus infected cells at a time when cap-dependent translation of cellular mRNAs is inhibited. Proc Natl Acad Sci USA 86: 5795–5799

Schneider RJ, Zhang Y (1993) Translational regulation in adenovirus infected cells. In: Ilan J (ed) Translational regulation of gene expression, vol 2. Plenum, New York, pp 227–250

Smith MR, Saramllo M, Liv L-L, Dever TE, Merrick WC, Kung HF and Sonenberg N (1990) Translation initiation factors induce DNA synthesis and transform NIH 3T3 cells. New Biol 2: 648–654

Sonenberg N, Guertin D, Lee KAW (1982) Capped mRNAs with reduced secondary structure can function in extracts from poliovirus infected cells. Mol Cell Biol 2: 1633–1638

Thach RE (1992) Cap recap: the involvement of eIF-4F in regulating gene expression. Cell 68: 177–180

Thimmapaya B, Ghadge GD, Rajan P, Swaminathan S (1993) Translational control by adenovirus-associated RNA I. In: Ilan J (ed) Translational regulation of gene expression, vol 2. Plenum, New York, pp 203–226

Thomas AAM, Ter Haar E, Wellink J, Voorma HO (1991) Cowpea mosaic virus middle component RNA contains a sequence that allows internal binding of ribosomes and that requires eukaryotic initiation factor 4F for optimal translation. J Virol 65: 2953–2959

Thomas AM, Scheper GC, Kleijn M, DeBoer M, Voorma HO (1992) Dependence of the adenovirus tripartite leader on the p220 subunit of eukaryotic initation factor 4F during in vitro translation. Eur J Biochem 207: 471–477

Wold WSM, Gooding LR (1991) Region E3 of adenovirus: a cassette of genes involved in host immunosurveillance and virus-cell interactions. Virology 184: 1–8

Zapata JM, Maroto FG, Sierra JM (1991) Inactivation of mRNA cap-binding protein complex in Drosophila melanogaster embryos under heat shock. J Biol Chem 266: 16007–16014

Zhang y, Schneider RJ (1993) Adenovirus inhibition of cellular protein synthesis and the preferential translation of late viral mRNAs. Semin Virol 4: 229–236

Zhang Y, Dolph PJ, Schneider RJ (1989) Secondary structure analysis of adenovirus tripartite leader. J Biol Chem 264: 10679–10684

Cap-Independent Translation of Heat Shock Messenger RNAs

R.E. Rhoads and B.J. Lamphear

1 Introduction

An early concept in the development of the field of translational control was that mRNAs differ in their intrinsic efficiencies of binding to ribosomes (Lodish 1976). Dozens of examples of mRNAs have now been described for which differences in efficiency of translation are attributed to differences in cap accessibility, secondary structure, sequence context of initiation codons, primary structure motifs,

Department of Biochemistry and Molecular Biology, Louisiana State University Medical Center, 1501 Kings Highway, Shreveport, LA 71130–3932, USA etc. Some of the first mRNAs to be studied in the context of translational control were those encoding the heat shock proteins (HSPs). Extensive investigations of the HSP mRNAs and the changes in the translational apparatus which accompany heat shock have provided one of the clearest examples of translational control. Despite these efforts, the mechanism(s) which direct the switchover from translation of non-HSP mRNAs to HSP mRNAs are not well understood. New insight into this problem has been provided by the discovery of cap-independent translation. The study of picornavirus infection, reviewed elsewhere in this volume, has led to the view that some mRNAs are initiated by a mechanism which is qualitatively different from that of most cellular mRNAs and which does not require the cap at all, much less the cellular machinery which recognizes the cap. It has now become apparent that there are numerous similarities between translation in heat-shocked cells and translation in picornavirus-infected cells and other cells which predominately carry out cap-independent translation. The evidence suggests that HSP mRNAs may be translated by the cap-independent route.

The purpose of the present article is to review the experimental results supporting the existence of a cap-independent state of the translational apparatus, the features of HSP mRNAs which enable them to be translated in this state, and the changes in the translational apparatus during heat shock which lead to this state. Space limitations do not permit a review of the extensive literature on the heat shock response itself, the biochemical functions and activities of HSPs, or the regulation of HSP production at the levels of transcription, mRNA splicing, or mRNA stability. The reader is referred to several excellent, recent reviews of these topics (Lindquist and Petersen 1990; Pain and Clemens 1990; Pauli et al. 1992; Sierra and Zapata 1994).

2 The Initiation of Translation

2.1 Cap-Dependent Translation

Figure 1 summarizes the various steps in the pathway of initiation as it is understood for capped mRNAs. The detailed series of events is covered in several recent reviews (Hershey 1991; Kozak 1992b; Merrick 1992; Thach 1992; Rhoads 1993). However, it is necessary to recapitulate a number of points which are relevant to the following discussion of the changes in the translational machinery caused by heat shock and other situations leading to the cap-independent state.

Upon dissociation of the two ribosomal subunits after a round of translation, the 40S subunit is converted to the lower density $43S_N$ subunit by binding of the large, multisubunit initiation factor eIF-3 (step 1). The 43S initiation complex is formed by addition of eIF-2, Met-tRNA$_i$, and GTP (step 2). This initiation complex binds mRNA at or near the cap and then scans until the first AUG in good

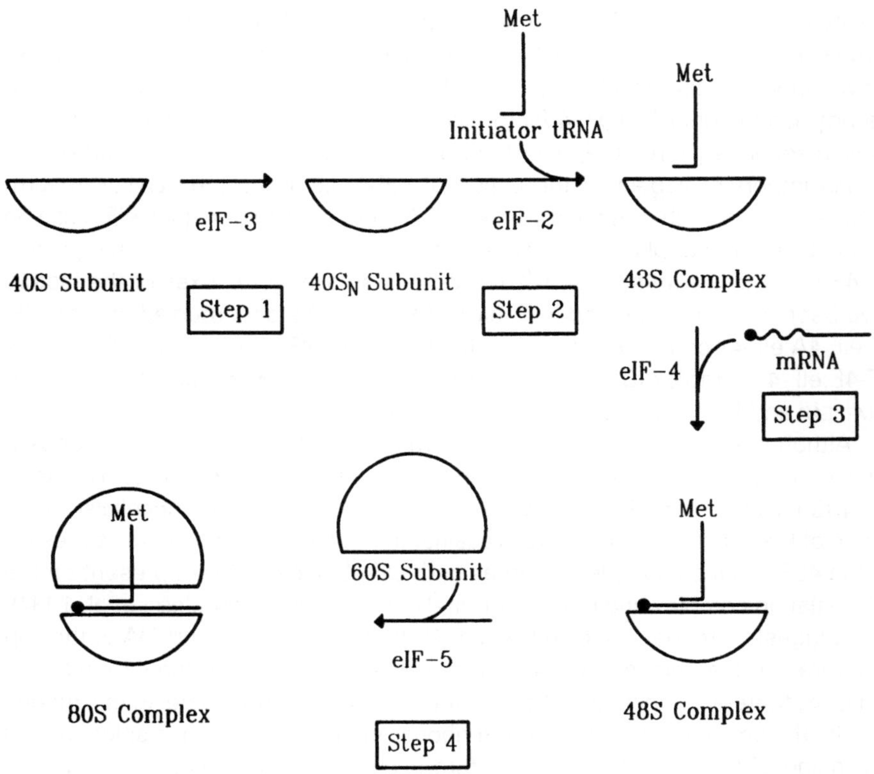

Fig. 1. Model for initiation of protein synthesis. The designation of eIF-2, eIF-3, elf-4 and eIF-5 as the catalysts for each step indicates, in some cases, classes of initiation factors. Factors which are involved in several steps, or whose stages of involvement are not known, are not listed. The *wavy line* at the left end of mRNA signifies 5'-terminal secondary structure. (From RHOADS 1993)

sequence context is encountered (step 3). The bound GTP is hydrolyzed, the initiation factors dissociate, the 60S ribosomal subunit joins, and the 80S initiation complex is ready for synthesis of the first peptide bond (step 4).

Step 2 is catalyzed by polypeptides in the eIF-2 group. eIF-2 itself is a heterotrimer consisting of α, β and γ subunits and forms a ternary complex with Met-$tRNA_i$ and GTP. At the end of the initiation cycle, eIF-2 is released with a tightly bound GDP. Exchange of the GDP for GTP is catalyzed by the action of the guanine nucleotide exchange factor eIF-2B. Regulation of this step can occur through phosphorylation of eIF-2α (discussed in Sect. 5.2).

The binding of mRNA to the 43S initiation complex (step 3) is catalyzed by polypeptides of the eIF-4 group and is another major site for regulation of the initiation rate. The eIF-4 factors are an interacting group of polypeptides consisting of eIF-4E, a 25 kDa cap-binding protein; eIF-4A, a 46 kDa ATP-dependent RNA helicase; eIF-4B, a 70 kDa RNA-binding protein that stimulates the helicase activity of eIF-4A; and eIF-4γ (formerly p220), a 154 kDa polypeptide that forms complexes with the other eIF-4 polypeptides but which does not have other known biochemical activities. (The molecular masses sited are those of the mammalian factors.) The eIF-4 factors collectively recognize the mRNA cap and unwind secondary structure, prior to or during scanning by the 40S subunit.

An important aspect of the action of eIF-4 factors is their ability to form complexes with each other and with eIF-3. All four eIF-4 factors plus eIF-3 can be bound to m^7GTP-Sepharose (Lamphear and Panniers 1990), even though only eIF-4E has intrinsic affinity for m^7GTP. Various smaller complexes of the factors have been isolated (reviewed in Rhoads 1991), but the best studied are the eIF-4E:eIF-4A:eIF-4γ complex, which has been termed eIF-4F (Safer 1989), and the eIF-4E:eIF-4γ complex (Buckley and Ehrenfeld 1987; Etchison and Milburn 1987; Lamphear and Panniers 1990; Zapata et al. 1994).

Although a number of investigators have proposed models for cap-dependent initiation in which the eIF-4F complex interacts with mRNA prior to its binding to the 43S initiation complex (reviewed in Joshi et al. 1994), the subcellular distribution of the initiation factors argues against this. eIF-4γ, but not eIF-4E, is present on the 43S initiation complex whereas both eIF-4E and eIF-4γ are present on the 48S initiation complex and in approximately 1:1 stoichiometry (Joshi et al. 1994). This suggests a model in which free eIF-4E first interacts with mRNA at the cap structure and then binds to eIF-4γ, which is already located on the 43S initiation complex, thereby bringing the 5' end of the mRNA to the ribosome. In this model, the eIF-4F complex assembles on the ribosome rather than free in solution, and unwinding of mRNA secondary structure occurs simultaneously with scanning rather than prior to ribosome binding. The regulated association of eIF-4E with eIF-4γ is particularly relevant to the heat-shock-induced inhibition of translation (see Sect. 5.3). The process of 48S initiation complex formation also requires eIF-4A, eIF-4B and ATP (Anthony and Merrick 1992), but it is not clear whether the interaction of these factors with mRNA occurs before or after mRNA becomes bound to the ribosome.

Phosphorylation of eIF-4B, eIF-4E and eIF-4γ is associated with increased rates of translation, and a large number of mitogens, tumor promoters, polypeptide growth factors and oncogenes increase both phosphorylation of these factors and protein synthesis (reviewed in Hershey 1991; Rhoads 1993). Molecular genetic evidence has also been presented for the importance of eIF-4E phosphorylation in controlling the initiation rate. Recent studies indicate that phosphorylation of eIF-4E increases its affinity for the cap by three- to fourfold (Minich et al. 1994).

2.2 Cap-Independent Translation

The concept that there is a separate cap-independent pathway for initiation is relatively new and is currently the subject of intense investigation, as attested to by the studies described elsewhere in this volume. Early indications that the cell could translate some mRNAs by a cap-independent pathway came from the observation that poliovirus infection inactivated the cap recognition machinery, shutting off host mRNA translation but permitting viral mRNA translation (TRACHSEL et al. 1980; TAHARA et al. 1981; ETCHISON et al. 1982). A mechanism was provided with the observation that picornavirus RNA undergoes internal initiation, a process whereby the 43S initiation complex initially binds mRNA at some point downstream of the cap (PELLETIER and SONENBERG 1988; JANG et al. 1989; BELSHAM and BRANGWYN 1990; KAMINSKI et al. 1990; KÜHN et al. 1990; BROWN et al. 1991), although the experimental support for such a mechanism has been questioned (KOZAK 1992a). This phenomenon has now also been observed with cowpea mosaic virus (THOMAS et al. 1991; VERVER et al. 1991), hepatitis C virus (TSUKIYAMA-KOHARA et al. 1992), the human heavy chain immunoglobulin-binding protein (GRP78/BiP) (MACEJAK and SARNOW 1991), and the product of the *Antennapedia* gene in *Drosophila* (OH et al. 1992). The specific region within the 5' NCR which is responsible for internal initiation is referred to as an internal ribosome entry site (IRES).

The role of initiation factors in cap-independent initiation is unclear. Proteolytic cleavage of eIF-4γ occurs during infection with some classes of picornavirus, coincident with the loss of the cell's ability to translate capped mRNAs (ETCHISON et al. 1982). The clear implication is that cap-independent translation does not require eIF-4γ. Yet eIF-4 has been shown to stimulate internal initiation (ANTHONY and MERRICK 1991; THOMAS et al. 1991). It has also been suggested that the cleavage products of eIF-4γ are required for internal initiation, based on the stimulatory effects of either expression of the poliovirus 2A protease in vivo (HAMBIDGE and SARNOW 1992) or the addition of rhinovirus 2A protease in vitro (LIEBIG et al. 1993). Alternatively, 2A protease may have a direct role in internal initiation; mutations in the poliovirus 5' NCR are suppressed by mutations in the protease 2A gene (MACADAM et al. 1994). Other cellular proteins, not previously identified as initiation factors, may also participate in internal initiation. A 52 kDa protein was identified on the basis of its ability to bind to the internal ribosome entry site of poliovirus (MEEROVITCH et al. 1989) and was subsequently shown to correct aberrant translation of poliovirus RNA in reticulocyte lysates (MEEROVITCH et al. 1993). This protein was found to be identical to the La antigen, an RNA polymerase III termination factor (MEEROVITCH et al. 1993). A 57–58 kDa polypeptide was similarly identified on the basis of its binding to the IRES elements of encephalomyocarditis virus (EMCV) (BOROVJAGIN et al. 1990; JANG and WIMMER 1990) and foot and mouth disease virus (FMDV) (LUZ and BECK 1990). It was subsequently shown to be the same as the polypyrimidine-binding protein involved in 3'-splice site selection and spliceosome assembly on pre-mRNA (HELLEN et al. 1993). A variety of other proteins which bind to the poliovirus 5' NCR have been identified (reviewed in MEEROVITCH and SONENBERG 1993). How these or other proteins compensate for the putative loss of activities upon proteolytic cleavage of eIF-4γ, however, is unknown.

3 Experimental Systems in Which Selective Translation of Heat Shock Protein mRNAs Is Observed

Three separate experimental treatments of cells result in remarkably similar alterations in translational specificity. In two of these cases, the nature of the defect is understood, at least partially. The overall similarity of the three altered translational states suggests that the defect in the third case is the same.

3.1 Heat Shock

The effects of heat shock on translational discrimination among mRNAs has been studied in a variety of cell types (reviewed in Pain and Clemens 1990). The best studied systems involve cells from *Drosophila* (Lindquist and Petersen 1990; Pauli et al. 1992; Sierra and Zapata 1994), HeLa (Duncan and Hershey 1985; De Benedetti and Baglioni 1986a), Ehrlich ascites tumor (Panniers and Henshaw 1984) and *Xenopus* oocytes (Bienz and Gurdon 1982). Though the magnitude of changes varies between systems, the major characteristics of the response are the same. The synthesis of normal (non-HSP) proteins rapidly decreases, the synthesis of HSPs remains the same or increases, polysomes disaggregate, non-HSP mRNAs shift to lower polysomes, and HSP mRNAs remain on higher polysomes. The change in translational specificity does not require transcription of HSP genes or the appearance of new HSPs in the cytosol. The non-HSP mRNAs are not degraded or inactivated during this time but are capable of being translated upon recovery. Return to normal temperatures restores the original rate and pattern of protein synthesis, and this recovery is achieved without synthesis of new RNA or protein. The defect produced by heat shock is preserved in cell lysates and can be corrected by the addition of protein factors (discussed in Sect. 5).

3.2 Picornavirus Infection

Picornavirus infection of mammalian cells also results in a drastic reduction in cellular protein synthesis, followed by a rise, after several hours, in the production of viral proteins (reviewed in Meerovitch et al. 1990). Cellular polysomes disaggregate and are replaced by virus-specific polysomes. Host mRNAs remain intact and unmodified, but these mRNAs are unable to bind 40S ribosomal subunits in poliovirus-infected cells. Although synthesis of virtually all cellular proteins decreases after poliovirus infection, the synthesis of HSP70 is more resistant (Muñoz et al. 1984). Another HSP, termed glucose-regulated protein 78 or immunoglobulin heavy-chain binding protein (GRP78/BiP), is actually synthesized at higher rates in poliovirus-infected cells (Sarnow 1989). The defect in the translational system is preserved in cellular extracts and can be corrected by the addition of protein factors (Trachsel et al. 1980; Tahara et al. 1981).

3.3 Depletion of eIF-4E and eIF-4γ with Antisense RNA

The third experimental situation in which HSP mRNAs are preferentially translated occurs when antisense RNA against eIF-4E mRNA is expressed in HeLa cells (subsequently referred to as AS cells; JOSHI-BARVE et al. 1992). An episomal vector was constructed from which 20 nucleotides of RNA complementary to the 5'-terminus of eIF-4E mRNA could be expressed under control of a dioxin-responsive enhancer (DE BENEDETTI and RHOADS 1991). Expression of antisense RNA caused the eIF-4E mRNA level to decrease 11-fold and the eIF-4E protein level to become undetectable (DE BENEDETTI et al. 1991). Depletion of eIF-4E caused protein synthesis to decrease drastically (Fig. 2). Addition of the inducer of antisense RNA expression, TCDD, did not affect protein synthesis in control cells or cells containing the vector without the antisense insert. Protein synthesis was depressed 2.8-fold in AS cells without induction and a further 4.3-fold after induction of antisense RNA.

Despite this decrease in overall protein synthesis, the steady-state levels of HSP72/73 and HSP65 remained constant and those of HSP90 and HSP27 actually increased. The rates of synthesis of HSP72/73 and HSP90 increased dramatically even though their mRNAs levels either remained constant or decreased

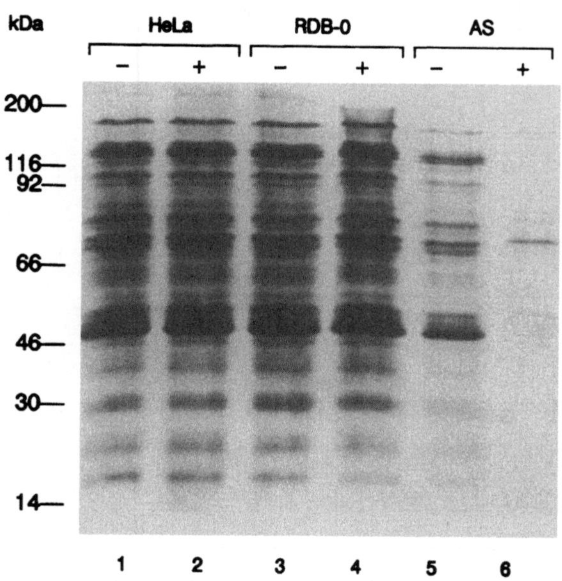

Fig. 2. Protein synthesis rates in control and HeLa cells expressing antisense RNA against eIF-4E mRNA (AS cells). Approximately 10^6 cells per sample were labeled for 3 h with 30 μCi/ml [3,4,5-^{3}H]-leucine. Cells, transformed with the vector alone (RDB-0) or the same vector expressing antisense RNA complementary to the first 20 nucleotides of eIF-4E mRNA (RDB-AS), were grown in 0.2 mg/ml G418. These, plus untransformed HeLa cells, were incubated with (+) and without (–) the inducer of antisense RNA expression, TCDD, for 48 h prior to labeling and were lysed and subjected to SDS-PAGE on 10% gels. (From DE BENEDETTI et al. 1991)

(Joshi-Barve et al. 1992). Polysomes were disaggregated (Fig. 3A), but non-HSP and HSP mRNAs behaved in opposite ways: actin mRNA shifted from high to low polysomes (Fig. 3B) while HSP90 mRNA shifted from low to high polysomes (Fig. 3C). The defect was preserved in extracts and could be corrected by the addition of protein factors (De Benedetti et al. 1991). Surprisingly, however, eIF-4E alone was unable to restore activity whereas the eIF-4 complex was. Direct determination of the levels of eIF-4 polypeptides revealed that eIF-4γ levels decreased simultaneously with eIF-4E upon induction of AS RNA, but eIF-4A levels remained constant (Fig. 4).

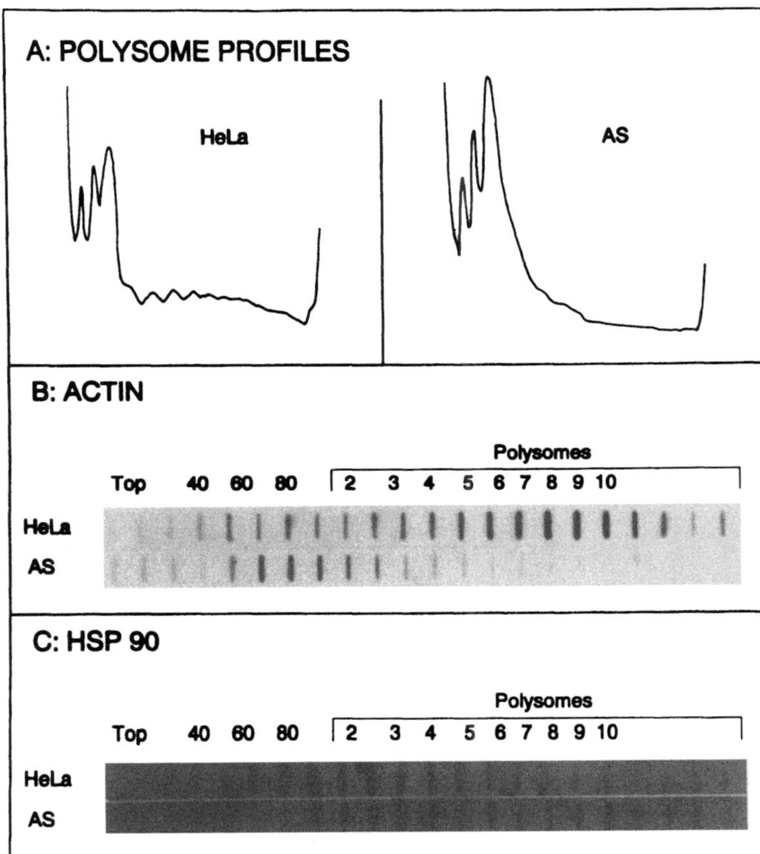

Fig. 3A–C. Analysis of polysomal mRNA distribution in HeLa cells expressing antisense RNA against eIF-4E mRNA (AS cells). Control HeLa cells and AS cells treated with the antisense inducer TCDD for 48 h were harvested and analyzed for polysomes. **A** Optical density profiles of polysomes. The direction of sedimentation is left to right. **B** The polysomal distribution of β-actin mRNA as determined by hybridization to actin cDNA using a slot blot apparatus. The ribosomal subunits are indicated by 40 and 60, monosomes by 80, and polysomes by 2, 3, (disome, trisome), etc. **C** Same as **B** except cDNA to HSP90 was used for hybridization. (From Joshi-Barve et al. 1992)

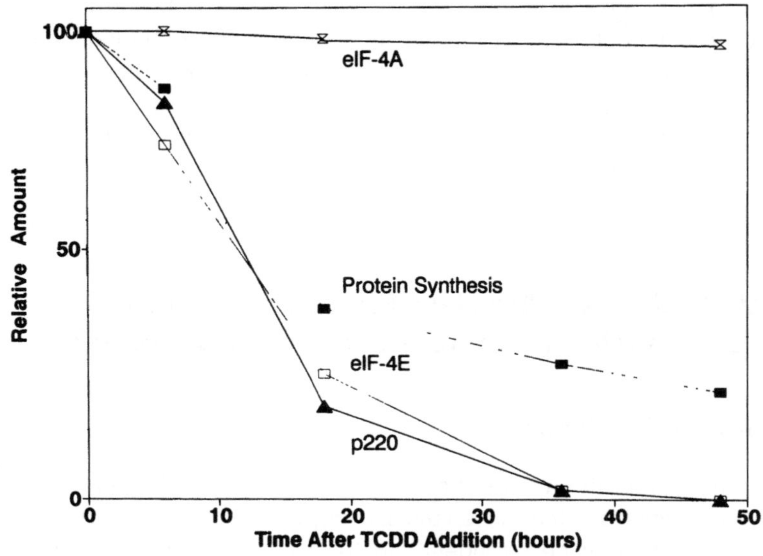

Fig. 4. Decay rates of eIF-4A, eIF-4E, eIF-4γ (p220), and protein synthesis in HeLa cells expressing antisense RNA against eIF-4E mRNA (AS cells). AS cells were incubated with the antisense inducer TCDD in multiwell plates for the times indicated. Protein synthesis rates were measured in cells which were pulse labeled for 3 h with [^{3}H]leucine (*closed squares*). eIF-4E was measured in cells labeled to equilibrium with [^{3}H]leucine for 48 h. The cells were lysed and eIF-4E was isolated by affinity chromatography, subjected to SDS-PAGE, and the radioactivity quantitated by fluorography (*open squares*). eIF-4γ (*closed triangles*) and eIF-4A (*hourglasses*) were measured by western blotting. All the values in this figure are expressed relative to AS cells not treated with TCDD. (From De Benedetti et al. 1991)

3.4 Similarities Among the Systems

In all three systems, heat shock, picornavirus infection, and expression of antisense RNA against eIF-4E, the characteristics of the translational system are qualitatively similar. Non-HSP protein synthesis is drastically decreased, HSP protein synthesis remains the same or increases, bulk polysomes are disaggregated, non-HSP mRNAs move from high to low polysomes while HSP mRNAs do the opposite, and the defect can be corrected in cell-free extracts by the addition of protein factors. In the picornavirus-infected cell, we can conclude that the translational machinery has switched over to a cap-independent state because: (1) picornavirus mRNA is uncapped; (2) initiation occurs by internal entry of the 43S initiation complex rather than by scanning from the 5'-terminus, and (3) one component of the cap-recognition apparatus, eIF-4γ, is proteolytically inactivated (although cleavage products remain). We can similarly conclude that the AS cell, when AS RNA expression is fully induced, is in a cap-independent state, since two components of the cap-recognition apparatus, eIF-4E and eIF-4γ, are completely absent (including eIF-4γ cleavage products). By analogy, we would surmise that the heat-shocked cell is similarly in the cap-independent state. Additional evidence for this conclusion is provided by an analysis of individual components of the translation system (Sect. 5). Furthermore, since HSP mRNAs are translated in all three systems, we would conclude that these mRNAs are capable of utilizing the cap-independent pathway.

4 By What Mechanism Are Heat Shock Protein mRNAs Able to Be Translated in Cap-Independent Systems?

4.1 The Competition Model

A competition model for mRNA translation states that each mRNA has a unique rate constant for initiation and that, when the capacity for translation is reduced, a competition results whereby the "strongest" mRNAs are selected most frequently by some critical component of the translational machinery (LODISH 1976). Several groups have suggested that HSP mRNAs simply outcompete the normal cellular messages for a limiting factor involved in translation (HICKEY and WEBER 1982; MUÑOZ et al. 1984; PANNIERS and HENSHAW 1984; PANNIERS et al. 1985; JACKSON 1986). It is clear that HSP mRNAs are efficient at initiating translation (LINDQUIST 1980; HICKEY and WEBER 1982; JACKSON 1986). However, simple competition for translational components is insufficient to explain all of the data, the evidence from *Drosophila* being particularly compelling. Typically in *Drosophila*, normal mRNA translation becomes inhibited before HSP mRNAs become abundant in the cytoplasm (LINDQUIST 1981), although this is not always the case (JACKSON 1986). In addition, this inhibition of translation can occur in the presence of actinomycin D, i.e., when HSP mRNA production has been blocked altogether (LINDQUIST 1980). This indicates that the translational inhibition is independent of the presence of HSP mRNAs and does not result from the synthesis of HSPs themselves. Furthermore, cotranslation of HSP mRNAs and normal mRNAs either in vitro or in vivo does not result in competition (STORTI et al. 1980; KRÜGER and BENECKE 1981; LINDQUIST 1981; SCOTT and PARDUE 1981). In addition, the coordinated fashion in which translation of all non-HSP mRNAs become modulated during heat shock and recovery (LINDQUIST 1980; STORTI et al. 1980; DIDOMENICO et al. 1982) is contrary to predictions of the Lodish theory, which is based on a continuous spectrum of mRNA efficiencies.

Further evidence against a competition model comes from studies with AS cells (JOSHI-BARVE et al. 1992). As shown in Fig. 3B, northern blot analysis of polysome profiles from AS cells indicates that actin, an mRNA that is translated efficiently in control cells, localizes with large polysomes in control cells (HeLa). However, in AS cells, when polysomes shrink in size and number, this typical mRNA sediments in the nonpolysomal and small polysomal regions of the gradient. HSP90 transcripts, which are utilized relatively inefficiently under normal conditions, localize to small polysomes in control cells but shift to large polysomes under AS conditions. This is inconsistent with the notion that HSP mRNAs compete more effectively for translational components, since this would predict that HSP mRNAs occupy large polysomes under control conditions as well. Also, a return to normal translation after heat shock is accompanied by a decrease in HSP synthesis without an appreciable decrease in HSP mRNA levels (De Benedetti and Baglioni 1986b). This suggests that HSP mRNAs are actually outcompeted for translation under normal conditions. This is at odds with the view that heat shock mRNAs are translated as "strong" messages.

4.2 Conserved Sequences in the 5' NCR of Heat Shock Protein mRNAs

The signal for preferential translation of HSP mRNAs during heat shock appears to be located within their 5'-NCRs (reviewed by Lindquist and Petersen 1990). There is an homologous region in the 5'-NCRs of *Drosophila* HSP22, 23, 26, 27, and 70 mRNAs (but notably absent in that of HSP 83) which begins near the 5'-terminus and extends for about 20 nts (Ingolia and Craig 1981; Hackett and Lis 1983). This region is necessary for translation under heat shock conditions (Hultmark et al. 1986). In addition, its location near the 5' terminus appears to be important since messages that contain an insertion that moves this sequence to an internal position within the 5'-NCR renders the message virtually untranslatable during heat shock (McGarry and Lindquist 1985). However, additional features of the leader contribute to translation during heat shock, since in the complete absence of this consensus sequence these mRNAs can still be translated during heat shock (McGarry and Lindquist 1985). A very recent review of 140 5'-NCRs of eukaryotic HSP mRNAs has revealed additional references which may participate in internal initiation (Joshi and Nguyen 1995).

4.3 Secondary Structure and Cap Dependence

The 5'-NCRs of *Drosophila* HSP mRNAs are longer than usual and are rich in A residues, a feature which should confer a low amount of secondary structure (Holmgren et al. 1981; Ingolia and Craig 1981; Hackett and Lis 1983; Hultmark et al. 1986). It is well known that mRNA secondary structure can have profound effects on mRNA translatability (reviewed in Pelletier and Sonenberg 1987). mRNAs with reduced secondary structure show less dependence on the cap, and their binding to ribosomes is resistant to inhibition by high salt, which stabilizes mRNA secondary structure (Herson et al. 1979; Edery et al. 1984). Translation of HSP mRNAs has been shown to be more resistant to high salt than non-HSP mRNAs

(Hickey and Weber 1982; Panniers et al. 1985; Jackson 1986; Maroto and Sierra 1988). However, reduced secondary structure does not completely explain HSP mRNA selection; addition of A-rich sequences to the 5'-NCR of an HSP70 mRNA lacking nts 2–205 is unable to restore its ability to be translated in heat-shocked *Drosophila* cells (Lindquist and Petersen 1990).

The requirement of a given mRNA for eIF-4 factors is proportional to the extent of 5'-NCR secondary structure (Sonenberg 1988; Koromilas et al. 1992). HSP mRNAs have reduced dependence on eIF-4 factors. Antibodies directed against either *Drosophila* eIF-4E (Zapata et al. 1991) or eIF-4γ (Zapata et al. 1994) inhibit translation of most non-HSP mRNAs but not of HSP mRNAs. In addition, lysates from heat-shocked cells preferentially translate HSP mRNAs in the presence of cap analogs, a condition which severely inhibits translation of normal capped mRNAs (Maroto and Sierra 1988). The lower dependence on eIF-4 factors thus provides HSP mRNAs with a selective advantage over non-HSP mRNAs under conditions in which the levels or activities of these factors are decreased.

The special nature of the leader sequences of HSP mRNAs is not confined to *Drosophila*. The 5'-NCR of the HSP GRP78/BiP directs internal initiation of translation, based on studies using bicistronic constructs (Macejak and Sarnow 1991). Also, the presence of the 5'-NCR of yeast HSP26 reduces cap dependence in a yeast cell-free system (Gerstel et al. 1992). In summary, the leader sequences of HSP mRNAs from various organisms contain information which is necessary for translation under heat shock conditions and which confers a reduced dependence on the cap and cap binding factors for translation.

5 Alterations in the Translational Apparatus During Heat Shock: Generation of a Cap-Independent State

What changes in the translational machinery result in the selective repression of non-HSP mRNAs and generation of a cap-independent state? Evidence from several laboratories suggests that in both *Drosophila* and mammalian systems the alteration occurs at the level of polypeptide chain initiation, although elongation may also be affected.

Progress in understanding this phenomenon has been greatly aided by the fact that the changes in translational selectivity observed at the cellular level are retained upon preparation of cell-free systems from heat-shocked cells (Storti et al. 1980; Kruger and Benecke 1981; Panniers et al. 1985; Maroto and Sierra 1988). Lysates from heat-shocked cells are impaired in their ability to translate non-HSP mRNAs yet retain the ability to translate HSP mRNAs. Restoration of normal patterns of translation can be obtained by supplementing these lysates with fractions obtained from normal cells. The fact that recovery of non-HSP mRNA translation upon return to normal temperatures does not require the synthesis of new RNA or protein molecules (Lindquist 1981; Muñoz et al. 1984;

PANNIERS and HENSHAW 1984) implies that the factors involved are not irreversibly inactivated during heat shock. The initial experiments suggested that the restoring activity was present in a crude ribosomal fraction (SCOTT and PARDUE 1981). However, a later report suggested the discriminatory factor was present in a soluble fraction (SANDERS et al. 1986). The reason for the discrepancy is unclear but may reflect differences in the fractionation conditions.

5.1 Ribosomal Protein S6

A number of candidates have been implicated as potential targets for reversible inactivation during heat shock. Dephosphorylation of ribosomal protein S6 has been observed in cultured tomato cells (SCHARF and NOVER 1982), *Drosophila* cells (GLOVER 1982), and mammalian cells (KENNEDY et al. 1984) in response to elevated temperatures. However, the role of dephosphorylation of S6 is unclear since the translational block occurs before dephosphorylation (TAS and MARTINI 1987), and no significant rephosphorylation of S6 occurs upon recovery of normal protein synthesis (OLSEN et al. 1983). In addition, chemical agents which induce HSP production and translational inhibition do not affect S6 phosphorylation levels (OLSEN et al. 1983).

5.2 Group 2 Translational Initiation Factors

A more promising candidate for alteration by heat shock is eIF-2. Phosphorylation of eIF-2α hinders the productive guanine nucleotide exchange necessary for translation initiation (reviewed in PAIN 1986; PROUD 1986). Changes in the level of phosphorylation of eIF-2α have been observed upon heat shock of HeLa cells (DUNCAN and HERSHEY 1984; DE BENEDETTI and BAGLIONI 1986a) and Ehrlich cells (SCORSONE et al. 1987). The activity of eIF-2 is reduced after heat shock (DUNCAN and HERSHEY 1984) as is guanine nucleotide exchange activity (ROWLANDS et al. 1988).

Despite these correlations between alterations in eIF-2 factors and translational inhibition, the evidence for a cause-and-effect relationship is contradictory. Addition of eIF-2 and eIF-2B to heat-shocked rabbit reticulocyte lysates stimulates protein synthesis (ERNST et al. 1982). However, in the better characterized systems of heat-shocked Ehrlich cells (PANNIERS et al. 1985) or *Drosophila* cells (ZAPATA et al. 1991), addition of eIF-2 to extracts does not stimulate translation, whereas eIF-4 does (see Sect. 5.3). Molecular genetic studies have shown that the expression of phosphorylation-resistant forms of eIF-2α in Chinese hamster ovary cells lessens the inhibition of protein synthesis that results from heat shock (MURTHA et al. 1993). This provides evidence that eIF-2α phosphorylation plays a role in the translational regulation during heat shock but suggests that additional points of regulation exist as well. Another indication of this is the fact that, under mild heat shock of HeLa cells, significant inhibition of protein synthesis occurs without a change in eIF-2α phosphorylation (DUNCAN and HERSHEY 1989).

The heat-shock induced phosphorylation of eIF-2α in HeLa cells appears to be due to the activity of an hemin-regulated inhibitor (HRI)-like kinase (De Benedetti and Baglioni 1986a). Although HRI was originally discovered in connection with the translational inhibition that accompanies heme deprivation in reticulocytes, other mammalian HRI-like kinases have been reported (Olmsted et al. 1993), suggesting a broader role of such enzymes in translational regulation. The mechanism of activation of the kinase in stress conditions has been the focus of recent attention (Matts and Hurst 1989, 1992; Mendez et al. 1992; Matts et al. 1992, 1993). HRI associates with several known HSPs, and the state of association is regulated by various cellular parameters. HRI coimmunoprecipitates with HSP90, HSP70, and a 56 kDa protein. Binding to HSP90 and p56 occurs only in the presence of hemin, and inhibition of reticulocyte lysates through eIF-2α phosphorylation by a number of stress conditions is inversely related to the level of HSP70 and p56. These data are consistent with the hypothesis that rising levels of denatured proteins that result from elevated temperatures and other stress conditions compete with HRI for association with HSP70; this causes dissociation of HSPs from the HRI and activation of the kinase. In support of this idea, a variety of proteins, when denatured and added to a reticulocyte lysate, inhibit protein synthesis, whereas the addition of the native proteins has no effect. Inhibition is reversed upon the addition of cAMP or eIF-2B. An antibody to HRI prevents the rise in eIF-2α phosphorylation caused by denatured protein, and this antibody fails to coimmunoprecipitate HSP70 in the presence of the denatured protein.

Despite the evidence that eIF-2α phosphorylation plays a role in the inhibition of protein synthesis during heat shock, it is unclear how this event would contribute to the cap-independent translation of HSP mRNAs. There is some evidence for a role for eIF-2 in mRNA discrimination (Thomas et al. 1992), supported in part by the fact that some viral 5'-NCRs appear to bind eIF-2/eIF-2B efficiently (Scheper et al. 1991) and therefore may aid in their selection. However, the bulk of the evidence on mRNA discrimination implicates the eIF-4 factors, which are required for the binding of mRNA to the 43S initiation complex, rather than the eIF-2 factors (reviewed in Rhoads 1991; Thach 1992).

5.3 Group 4 Translational Initiation Factors

Analysis of eIF-4 factors has yielded much information on potential targets for regulation during heat shock. A considerable amount of evidence links phosphorylation of eIF-4 factors to their increased activity (see Sect. 2.1). The effect of heat shock on both the phosphorylation and the activity of eIF-4 factors has been studied.

Both eIF-4B and eIF-4E are dephosphorylated in response to heat shock (Duncan et al. 1987). Dephosphorylation of eIF-4E occurs for both the free form as well as the eIF-4γ-associated form of the factor (Lamphear and Panniers 1991). Since phosphorylation of eIF-4E increases its affinity for caps (Minich et al. 1994), the dephosphorylation of eIF-4E would be expected to be unfavorable for mRNAs translated by the cap-dependent route, but not to affect mRNAs translated by the cap-independent route. Some doubt is cast on this mechanism, however, by the findings that eIF-4E dephosphorylation does not occur under mild heat stress conditions and that rephosphorylation of eIF-4B to pre-stress levels can occur prior to recovery of protein synthesis (DUNCAN and HERSHEY 1989). Regardless of the uncertainty about the role of phosphorylation of eIF-4 factors, it is clear that heat shock alters the activity of factors involved in cap recognition. The activity of an eIF-4F/eIF-3 fraction from HeLa cells is reduced after heat shock (DUNCAN and HERSHEY 1984). Furthermore, the translation of normal messages in lysates prepared from heat-shocked Ehrlich cells is preferentially stimulated by the addition of rabbit reticulocyte eIF-4F (PANNIERS et al. 1985). The restoring factor was purified from non-heat-shocked Ehrlich cell extracts by m^7GTP-Sepharose and Mono Q chromatography (LAMPHEAR and PANNIERS 1990). Figure 5 shows that an eIF-4E:eIF-4γ complex is obtained and that this complex cofractionates with restoring activity. The activity is clearly separated from eIF-4A, eIF-4B, and eIF-3. The inhibition of translation seen in fractions containing free eIF-4E is due to the presence of m^7GTP which elutes from Mono Q with eIF-4E. Extensive dialysis reveals a small amount of restoring activity is also associated with these fractions (data not shown). However, comparison of restoring activity present in preparations of eIF-4E with that of the eIF-4E:eIF-4γ complex indicates that eIF-4E is much less efficient on a molar basis at restoring translation (LAMPHEAR and PANNIERS 1990). *Drosophila* heat shock lysates were shown to be stimulated in the same manner by a complex containing eIF-4E and eIF-4γ, but free eIF-4E was relatively ineffective (ZAPATA et al. 1991).

From the foregoing, it is clear that heat shock reduces both the phosphorylation and the activity of eIF-4 polypeptides. It is tempting to speculate that this represents cause and effect, and that the translational stimulation by the Ehrlich cell complex containing highly phosphorylated eIF-4E is simply due to replenishment of phosphorylated eIF-4E. This does not seem to be the case, however, since free eIF-4E fractions which are nearly 50% phosphorylated are only about 5% as potent as the eIF-4E:eIF-4γ complex (LAMPHEAR and PANNIERS 1990). Another possible mechanism is that lysates from heat-shocked cells have reduced capacity to form eIF-4E:eIF-4γ complexes. Heat-shocked Ehrlich cell lysates have reduced restoring activity (LAMPHEAR and PANNIERS 1991), and less eIF-4γ is retained on m^7GTP-Sepharose columns with extracts from heat-shocked Ehrlich (LAMPHEAR and PANNIERS 1991), HeLa (DUNCAN et al. 1987), and *Drosophila* (ZAPATA et al. 1991) cells. Figure 6 shows the cap-binding proteins from control and heat shocked Ehrlich cells fractionated by Mono Q chromatography. Less eIF-4E:eIF-4γ complex is obtained from heat-shocked cells whereas there is little change in the recovery of eIF-4E. One interpretation is that heat shock prevents formation of the eIF-4E:eIF-4γ complex, perhaps through denaturation of eIF-4γ. Alternatively, the decreased phosphorylation of eIF-4E or eIF-4γ may impair complex formation.

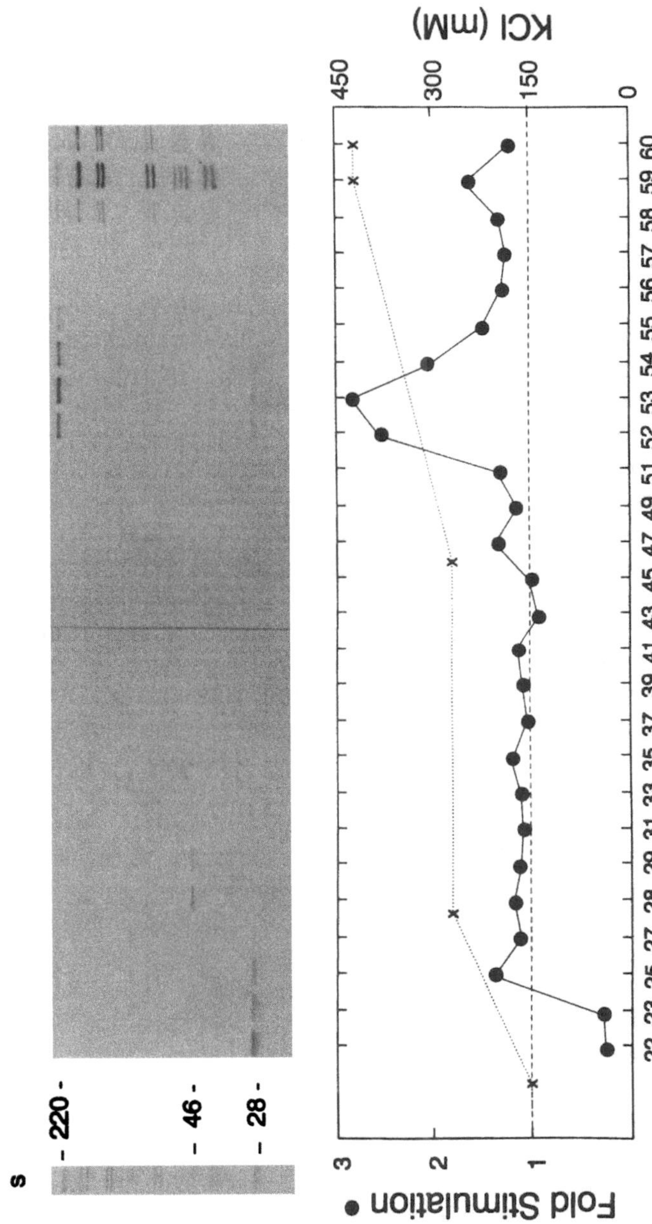

Fig. 5. Purification of restoring activity to extracts of heat-shocked Ehrlich cells by chromatography on Mono Q. The high salt ribosomal wash from non-heat-shocked Ehrlich cells was purified by m^7GTP-Sepharose, and the bound fraction (60 µg) was used as a starting material (*lane S*) for chromatography on Mono Q. The *upper panel* shows Coomassie stained SDS-PAGE analysis of relevant fractions (150 µl). The *lower panel* shows fold stimulation of protein synthesis in a lysate from heat-shocked Ehrlich cells for the same fractions (20 µl). The start and end points of the KCl gradient are indicated (*X*). (From LAMPHEAR and PANNIERS 1990)

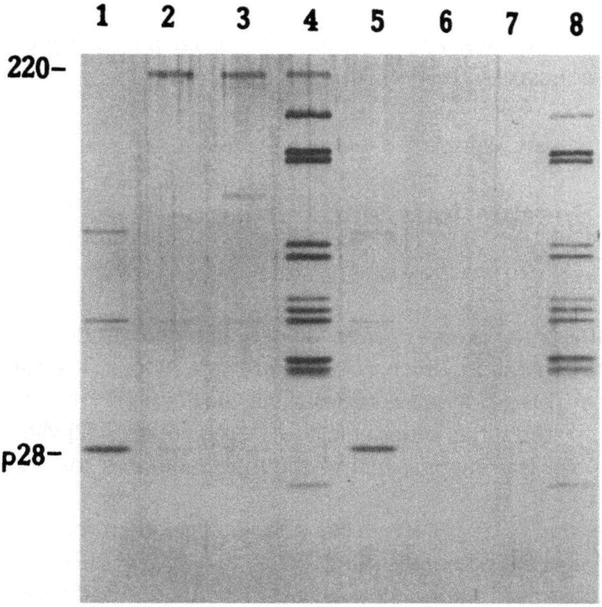

Fig. 6. Analysis of eIF-4E:eIF-4γ from control and heat-shocked Ehrlich cells by Mono Q chromatography and SDS-PAGE. Equal amounts of protein from the high salt post-ribosomal supernatants of control and heat-shocked Ehrlich cells were purified by m^{7}GTP-Sepharose chromatography, and the bound fractions (cap-binding proteins) were chromatographed on Mono Q columns as shown in Fig. 5. Aliquots of peak fractions (20 μl) from control (*lanes 1–4*) and heat-shocked (*lanes 5–8*) cell supernatants were analyzed by SDS-PAGE. Proteins were visualized with silver stain. *Lanes 1 and 5* represent the peak of free eIF-4E; *lanes 2 and 6* and *lanes 3 and 7* represent the first half and second half of the eIF-4E:eIF-4γ peak, respectively. *Lanes 4 and 8* represent the eIF-3 peak. Mobilities corresponding to 28 kDa and 220 kDa are indicated. (From LAMPHEAR and PANNIERS 1991)

5.4 Other Factors Involved in Protein Synthesis

Other elements of the protein synthesis apparatus may be involved in the alteration of translation caused by heat shock, but their roles are not understood at present. HSP70 proteins themselves may participate in protein synthesis. In yeast the HSP70 proteins SSB are associated with active polysomes, and mutant ssb1 and ssb2 strains have a slow growth phenotype which can be suppressed by increased copy number of a gene coding for an elongation factor 1α-like protein (NELSON et al. 1992). Furthermore, a yeast DnaJ homologue, SIS1, appears to play a role in translation initiation, since temperature-sensitive *sis1* strains rapidly accumulate 80S ribosomes and have a reduced number of polysomes (ZHONG and ARNDT 1993). This phenotype can be suppressed by alterations in the 60S ribosome. If HSP proteins are, in fact, required for normal translation, the following model might explain the inhibition of protein synthesis caused by heat shock: heat shock would lead to the generation of denatured proteins; these would bind to and sequester HSPs; the HSPs would not be available to participate in protein synthesis, thereby causing inhibition. Recovery of translation would occur when new HSPs were synthesized or when denatured proteins were renatured, freeing up the sequestered HSPs.

6 Open Questions, Speculations, and Future Directions

Despite the considerable amount of progress made to date in understanding the dramatic alteration in mRNA selection which occurs upon heat shock, the two central questions discussed in Sect. 4 and 5 remain unanswered: (1) How are HSP mRNAs able to be translated when other cellular mRNAs are not? (2) What is the reversible change in the translational machinery that generates the cap-independent state?

Although it seems clear that HSP mRNAs utilize a cap-independent route for initiation, the mechanism is not yet understood. Only in the case of GRP78/BiP has internal initiation been demonstrated; similar studies have not yet been performed for other HSP mRNAs. It should be noted, however, that the 5'-NCR of HSP mRNAs is considerably shorter than that of picornaviral RNAs which contain a well defined internal ribosome entry site. HSP mRNAs may utilize a mechanism of cap-independent initiation other than internal initiation. It is noteworthy that the adenovirus late mRNA tripartite leader confers cap-independent translation but does not promote internal initiation (Dolph et al. 1990).

The evidence reviewed in Sect. 5 indicates that both eIF-2 and eIF-4 undergo some degree of alteration during heat shock. Considering the importance of protein synthesis to cell survival, it is likely that there are multiple targets for its regulation. Recent investigation provides insight into a potential mechanism for the activation of an eIF-2α kinase during heat shock. Clearly eIF-4 activity is regulated as well. If eIF-4γ is indeed a component sensitive to heat shock, it remains to be determined what causes the inactivation; two possibilities are a change in phosphorylation and denaturation of the polypeptide. It is also unknown what aspect of eIF-4γ's role is affected, some possibilities being its binding to the 43S initiation complex, its binding to eIF-4E, and its participation in the unwinding of mRNA secondary structure catalyzed by eIF-4A and eIF-4B. The fact that HSP mRNAs can be translated in the absence of eIF-4E and eIF-4γ does not rule out the possibility that these components may play a modified role in HSP mRNA translation during heat shock. It is tempting to speculate that eIF-4γ loses its ability to bind eIF-4E and thereby to attach capped mRNA to the 43S initiation complex (according to the model presented in Sec. 2) but retains its ability to participate in unwinding. Such a bifunctional role for eIF-4γ might explain the apparently confusing observations, cited in Sect. 2, that eIF-4γ is cleaved in picornavirus infection but that it, or possibly its cleavage products, promote cap-independent initiation.

The subject of recovery of the translational apparatus from heat shock is also poorly understood. If one or more activities of eIF-4γ are lost upon heat shock, there may be a role of HSP70 or other HSPs in renaturing eIF-4γ, akin to the demonstrated ability of the *E. coli* HSP70 (Dnak) to renature RNA polymerase (SKOWYRA et al. 1990). It is interesting that, in AS cells, which are unable to accumulate detectable amounts of eIF-4γ, the translational apparatus is unable to recover from heat shock (JOSHI-BARVE et al. 1992).

Acknowledgments. We thank Dr. Jose M. Sierra for making information available prior to publication and Drs. Raul Mendez and Brett Keiper for helpful discussions and assistance with the manuscript. This study was supported by grant no. GM20818 from the National Institute of General Medical Sciences and grant no. 3076 from the Council for Tobacco Research-U.S.A., Inc.

References

Anthony DD, Merrick WC (1991) Eukaryotic initiation factor (eIF)-4F. Implications for a role in internal initiation of translation. J Biol Chem 266: 10218–10226
Anthony DD, Merrick WC (1992) Analysis of 40 S and 80 S complexes with mRNA as measured by sucrose density gradients and primer extension inhibition. J Biol Chem 267: 1554–1562
Belsham GJ, Brangwyn JK (1990) A region of the 5' noncoding region of foot-and-mouth disease virus RNA directs efficient internal initiation of protein synthesis within cells: involvement with the role of L protease in translational control. J Virol 64: 5389–5395
Bienz M, Gurdon J (1982) The heat shock response in *Xenopus* oocytes is controlled at the translational level. Cell 29: 811–819
Borovjagin AV, Evstafieva AG, Ugarova TY, Shatsky IN (1990) A factor that specifically binds to the 5'-untranslated region of encephalomyocarditis virus RNA. FEBS Lett 261: 237–240
Brown E, Day S, Jansen R, Lemon S (1991) The 5' nontranslated region of hepatitis A virus RNA: secondary structure and elements required for translation in vitro J Virol 65: 5828–5838
Buckley B, Ehrenfeld E (1987) The cap-binding protein complex in uninfected and poliovirus-infected HeLa cells. J Biol Chem 262: 13599–13606
De Benedetti A, Baglioni C (1986a) Activation of hemin-regulated initiation factor-2 kinase in heat-shocked HeLa cells. J Biol Chem 261: 338–342
De Benedetti A, Baglioni C (1986b) Translational regulation of the synthesis of a major heat shock protein in HeLa cells. J Biol Chem 261: 15800–15804
De Benedetti A, Rhoads RE (1991) A novel BK virus-based episomal vector for expression of foreign genes in mammalian cells. Nucleic Acids Res 19:1925–1931
De Benedetti A, Joshi-Barve S, Rinker-Schaeffer C, Rhoads RE (1991) Expression of antisense RNA against initiation factor eIF-4E mRNA in HeLa cells results in lengthened cell division times, diminished translation rates, and reduced levels of both eIF-4E and the p220 component of eIF-4F. Mol Cell Biol 11: 5435–5445
DiDomenico BJ, Bugaisky GE, Lindquist S (1982) Heat shock and recovery are mediated by different translational mechanisms. Proc Natl Acad Sci USA 79: 6181–6185
Dolph P, Huang J, Schneider R (1990) Translation by the adenovirus tripartite leader: Elements which determine independence from cap-binding protein complex. J Virol 64: 2669–2677
Duncan R, Hershey JWB (1984) Heat shock-induced translational alterations in HeLa cells. Initiation factor modifications and the inhibition of translation. J Biol Chem 259: 11882–11889
Duncan R, Hershey JWB (1985) Regulation of initiation factors during translational repression caused by serum depletion. Covalent modification. J Biol Chem 260: 5493–5497
Duncan RF, Hershey JWB (1989) Protein synthesis and protein phosphorylation during heat stress, recovery, and adaptation. J Cell Biol 109: 1467–1481
Duncan R, Milburn SC, Hershey JWB (1987) Regulated phosphorylation and low abundance of HeLa cell initiation factor eIF-4F suggest a role in translational control. J Biol Chem 262: 380–388

Edery I, Lee KAW, Sonenberg N (1984) Functional characterization of eukaryotic mRNA cap binding protein complex: effects on translation of capped and naturally uncapped RNAs. Biochemistry 23: 2456–2462

Ernst V, Baum EZ, Reddy P (1982) Heat shock phosphorylation and the control of translation in rabbit reticulocytes, reticulocytes lysates, and HeLa cells. In: Schlesinger M, Ashburner M, Tissieres A (eds) Heat shock from bacteria to man. Cold Spring Harbor Laboratory, Cold Spring Harbor, pp 215–225

Etchison D, Milburn S (1987) Separation of protein synthesis initiation factor eIF-4A from a p220-associated cap binding complex activity. Mol Cell Biochem 76: 15–25

Etchison D, Milburn SC, Edery I, Sonenberg N, Hershey JWB (1982) Inhibition of HeLa cell protein synthesis following poliovirus infection correlates with the proteolysis of a 220,000-dalton polypeptide associated with eucaryotic initiation factor 3 and a cap binding protein complex. J Biol Chem 257: 14806–14810

Gerstel B, Tuite MF, McCarthy JE (1992) The effects of 5' capping, 3' polyadenylation and leader composition upon the translational stability of mRNA in a cell-free extract derived from the yeast *Saccharomyces cerevisiae*. Mol Microbiol 6: 2339–2348

Glover CVC (1982) Heat shock induces rapid dephosphorylation of ribosomal protein in *Drosophila*. Proc Natl Acad Sci USA 79: 1781–1785

Hackett RW, Lis JT (1983) Localization of the hsp83 transcript within a 3292 nucleotide sequence from the 63B heat shock locus of *D. melanogaster* Nucleic Acids Res 11: 7011–7016

Hambidge SJ, Sarnow P (1992) Translational enhancement of the poliovirus 5' noncoding region mediated by virus-encoded polypeptide-2A. Proc Natl Acad Sci USA 89: 10272–10276

Hellen CUT, Witherell GW, Schmid M, Shin SH, Pestova TV, Gil A, Wimmer E (1993) A cytoplasmic 57-kDa protein that is required for translation of picornavirus RNA by internal ribosomal entry is identical to the nuclear pyrimidine tract-binding protein. Proc Natl Acad Sci USA 90: 7642–7646

Hershey JWB (1991) Translational control in mammalian cells. Annu Rev Biochem 60: 717–755

Herson D, Schmidt A, Seal SN, Marcus A, van Vloten-Doting L (1979) Competitive mRNA translation in an in vitro system from wheat germ. J Biol Chem 254: 8245–8249

Hickey ED, Weber LA (1982) Modulation of heat-shock polypeptide synthesis in HeLa cells during hyperthermia and recovery. Biochemistry 21: 1513–1521

Holmgren R, Corces V, Morimoto R, Blackman R, Meselson M (1981) Sequence homologies in the 5' regions of four *Drosophila* heat-shock genes. Proc Natl Acad Sci USA 78: 3775–3778

Hultmark D, Klemenz R, Gehring WJ (1986) Translational and transcriptional control elements in the untranslated leader of the heat-shock gene hsp22. Cell 44: 429–438

Ingolia TD, Craig EA (1981) Primary sequence of the 5' flanking regions of the *Drosophila* heat shock genes in chromosome subdivision 67B. Nucleic Acids Res 9: 1627–1642

Jackson RJ (1986) The heat-shock response in *Drosophila* KC 161 cells. Eur J Biochem 158: 623–634

Jang SK, Wimmer E (1990) Cap-independent translation of encephalomyocarditis virus RNA: structural elements of the internal ribosomal entry site and involvement of a cellular 57-kD RNA-binding protein. Genes Dev 4: 1560–1572

Jang SK, Davies M, Kaufman RJ, Wimmer E (1989) Initiation of protein synthesis by internal entry of ribosomes into the 5' nontranslated region of encephalomyocarditis virus RNA in vivo. J Virol 63: 1651–1660

Joshi B, Yan R, Rhoads RE (1994) In vitro synthesis of human protein synthesis initiation factor eIF-4γ and its localization on 43 S and 48 S initiation complexes. J Biol Chem 269: 2048–2055

Joshi CP, Nguyen HT (1995) Nucleic Acids Res 23: 541–549

Joshi-Barve S, De Benedetti A, Rhoads RE (1992) Preferential translation of heat shock mRNAs in HeLa cells deficient in protein synthesis initiation factors eIF-4E and eIF-4γ. J Biol Chem 267: 21038–21043

Kaminski A, Howell MT, Jackson RJ (1990) Initiation of encephalomyocarditis virus RNA translation: the authentic initiation site is not selected by a scanning mechanism. EMBO J 9: 3753–3759

Kennedy IM, Burdon RH, Leader DP (1984) Heat shock causes diverse changes in the phosphorylation of the ribosomal proteins of mammalian cells. FEBS Lett 169: 267–273

Koromilas AE, Lazaras-Karatzas A, Sonenberg N (1992) mRNAs containing extensive secondary structure in their 5' non-coding region translate efficiently in cells overexpressing initiation factor eIF-4E. EMBO J 11: 4153–4158

Kozak M (1992a) A consideration of alternative models for the initiation of translation in eukaryotes. CRC Crit Rev Biochem 27: 385–402

Kozak M (1992b) Regulation of translation in eukaryotic systems. Annu Rev Cell Biol 8: 197–225

Krüger C, Benecke BJ (1981) In vitro translation of *Drosophila* heat-shock and non-heat-shock mRNAs in heterologous and homologous cell-free systems. Cell 23: 595–603

Kühn R, Luz N, Beck E (1990) Functional analysis of the internal translation initiation site of foot-and-mouth disease virus. J Virol 64: 4625–4631

Lamphear BJ, Panniers R (1990) Cap binding protein complex that restores protein synthesis in heat-shocked Ehrlich cell lysates contains highly phosphorylated eIF-4E. J Biol Chem 265: 5333–5336

Lamphear BJ, Panniers R (1991) Heat shock impairs the interaction of cap-binding protein complex with 5' mRNA cap. J Biol Chem 266: 2789–2794

Liebig H-D, Ziegler E, Yan R, Hartmuth K, Klump H, Kowalski H, Blass D, Sommergruber W, Frasel L, Lamphear B, Rhoads RE, Kuechler E, Skern T (1993) Purification of two picornaviral 2A proteinases: interaction with eIF-4γ and influence on in vitro translation. Biochemistry 32: 7581–7588

Lindquist S (1980) Translational efficiency of heat-induced messages in *Drosophila melanogaster* cells. J Mol Biol 137: 151–158

Lindquist S (1981) Regulation of protein synthesis during heat shock. Nature 293: 311–314

Lindquist S, Petersen R (1990) Selective translation and degradation of heat-shock messenger RNAs in *Drosophila*. Enzyme 44: 147–166

Lodish HF (1976) Translational control of protein synthesis. Annu Rev Biochem 45: 39–72

Luz N, Beck E (1990) A cellular 57 kDa protein binds to two regions of the internal translation site of foot-and-mouth disease virus. FEBS Lett. 269: 311–314

Macadam A, Ferguson G, Fleming T, Stone D, Almond J, Minor P (1994) Role for poliovirus protease 2A in cap independent translation. EMBO J 13: 924–927

Macejak DG, Sarnow P (1991) Internal initiation of translation mediated by the 5' leader of a cellular mRNA. Nature 353: 90–94

Maroto FG, Sierra JM (1988) Translational control in heat-shocked *Drosophila* embryos J Biol Chem 263: 15720–15725

Matts RL, Hurst R (1989) Evidence for the association of the heme-regulated eIF-2α kinase with the 90-kDa heat shock protein in rabbit reticulocyte lysate in situ. J Biol Chem 264: 15542-15547

Matts RL, Hurst R (1992) The relationship between protein synthesis and heat shock proteins levels in rabbit reticulocyte lysates. J Biol Chem 267: 18168–18174

Matts RL, Xu ZY, Pal JK, Chen JJ (1992) Interactions of the heme-regulated eIF-2α kinase with heat shock proteins in rabbit reticulocyte lysates. J Biol Chem 267: 18160–18167

Matts RL, Hurst R, Xu Z (1993) Denatured proteins inhibit translation in hemin-supplemented rabbit reticulocyte lysate by inducing the activation of the heme-regualted eIF-2α kinase. Biochemistry 32: 7323–7328

McGarry TJ, Lindquist S (1985) The preferential translation of *Drosophila* hsp70 mRNA requires sequences in the untranslated leader. Cell 42: 903–911

Meerovitch K, Sonenberg N (1993) Internal initiation of picornavirus RNA translation. Semin Virol 4: 217–227

Meerovtich K, Pelletier J, Sonenberg N (1989) A cellular protein that binds to the 5'-noncoding region of poliovirus RNA: implications for internal translation initiation. Genes Dev. 3: 1026–1034

Meerovitch K, Sonenberg N, Pelletier J (1990) The translation of picornaviruses. In: Trachsel H (ed) Translation in eukaryotes. CRC Press, Berne, Switzerland, pp 273–292

Meerovitch K, Svitkin YV, Lee HS, Lejbkowicz F, Kenan DJ, Chan EKL, Agol VI, Keene JD, Sonenberg N (1993) La autoantigen enhances and corrects aberrant translation of poliovirus RNA in reticulocyte lysate. J Virol 67: 3798–3807

Mendez R, Moreno A, DeHaro C (1992) Regulation of heme-controlled eukaryotic polypeptide chain initiation factor-2α-subunit kinase of reticulocyte lysates. J Biol Chem 267: 11500–11507

Merrick WC (1992) Mechanism and regulation of eukaryotic protein synthesis. Microbiol Rev 56:291–315

Minich WB, Balasta ML, Goss DJ, Rhoads RE (1994) Chromatographic resolution of in vivo phosphorylated and nonphosphorylated translation initiation factor eIF-4E. Demonstration of increased cap affinity of the phosphorylated form. Proc Natl Acad Sci USA 91: 7668–7672

Muñoz A, Alonso MA, Carrasco L (1984) Synthesis of heat-shock proteins in HeLa cells: Inhibition by virus infection. Virology 137: 150–159

Murtha RP, Davies MV, Scherer BJ, Choi SY, Hershey JW, Kaufman RJ (1993) Expression of a phosphorylation-resistant eukaryotic initiation factor 2α-subunit mitigates heat shock inhibition of protein synthesis. J Biol Chem 268: 12946–12951

Nelson RJ, Ziegelhoffer T, Nicolet C, Werner-Washburne M, Craig EA (1992) The translation machinery and 70 kd heat shock protein cooperate in protein synthesis. Cell 71: 97–105

Oh SK, Scott MP, Sarnow P (1992) Homeotic gene *Antennapedia* messenger RNA contains 5'-noncoding sequences that confer translational initiation by internal ribosome binding. Genes Dev 6: 1643–1653

Olmsted EA, O'Brien L, Henshaw EC, Panniers R (1993) Purification and characterization of eukaryotic initiation factor (eIF)-2α kinases from Ehrlich ascites tumor cells. J Biol Chem 268: 12552–12559
Olsen AS, Triemer DF, Sanders MM (1983) Dephosphorylation of S6 and expression of the heat shock response in *Drosophila melanogaster*. Mol Cell Biol 3: 2017–2027
Pain V, Clemens M (1990) Adjustment of translation to special physiological conditions. In: Trachsel H (ed) Translation in eukaryotes. CRC Press, Berne, Switzerland, pp 293–324
Pain VM (1986) Initiation of protein synthesis in mammalian cells. Biochem J 235: 625–637
Panniers R, Henshaw EC (1984) Mechanism of inhibition of polypeptide chain initiation in heat-shocked Ehrlich ascites tumour cells. Eur J Biochem 140: 209–214
Panniers R, Stewart EB, Merrick WC, Henshaw EC (1985) Mechanism of inhibition of polypeptide chain initiation in heat-shocked Ehrlich cells involves reduction of eukaryotic initiation factor 4F activity. J Biol Chem 260: 9648–9653
Pauli D, Arrigo AP, Tissieres A (1992) Heat shock response in *Drosophila*. Experientia 48: 623–629
Pelletier J, Sonenberg N (1987) The involvement of mRNA secondary structure in protein synthesis. Biochem Cell Biol 65: 576–581
Pelletier J, Sonenberg N (1988) Internal initiation of translation of eukaryotic mRNA directed by a sequence derived from poliovirus RNA. Nature 334: 320–325
Proud CG (1986) Guanine nucleotides, protein phosphorylation and the control of translation. Trends Biochem Sci 11: 73–77
Rhoads RE (1991) Protein synthesis, cell growth and oncogenesis. Curr Biol 3: 1019–1024
Rhoads RE (1993) Regulation of eukaryotic protein synthesis by initiation factors. J Biol Chem 268: 3017–3020
Rowlands AG, Montine KS, Henshaw EC, Panniers R (1988) Physiological stresses inhibit guanine-nucleotide exchange factor in Ehrlich cells. Eur J Biochem 175: 93–99
Safer B (1989) Nomenclature of initiation, elongation and termination factors for translation in eukaryotes. Eur J Biochem 186: 1–3
Sanders MM, Triemer DF, Olsen AS (1986) Regulation of protein synthesis in heat-shocked Drosophila cells. Soluble factors control translation in vitro. J Biol Chem 261: 2189–2196
Sarnow P (1989) Translation of glucose-regulated protein 78/immunoglobulin heavy-chain binding protein mRNA is increased in poliovirus-infected cells at a time when cap-dependent translation of cellular mRNAs is inhibited. Proc Natl Acad Sci USA 86: 5795–5799
Scharf KD, Nover L (1982) Heat-shock-induced alterations of ribosomal protein phosphorylation in plant cell cultures. Cell 30: 427–437
Scheper GC, Thomas AAM, Voorma HO (1991) The 5' untranslated region of encephalomyocarditis virus contains a sequence for very efficient binding of eukaryotic initiation factor eIF-2/2B. Biochim Biophys Acta 1089: 220–226
Scorsone KA, Panniers R, Rowlands AG, Henshaw EC (1987) Phosphorylation of eukaryotic initiation factor 2 during physiological stresses which affect protein synthesis. J Biol Chem 262: 14538–14543
Scott MP, Pardue ML (1981) Translational control in lysates of Drosophila melanogaster cells. Proc Natl Acad Sci USA 78: 3353–3357
Sierra JM, Zapata JM (1994) Translational regulation of the heat shock response. Mol Biol Rep 19: 211–220
Skowyra D, Georgopoulos C, Zylicz M (1990) The *E. coli dnaK* gene product, the hsp70 homolog, can reactivate heat-inactivated RNA polymerase in an ATP hydrolysis-dependent manner. Cell 62: 939–944
Sonenberg N (1988) Cap-binding proteins of eukaryotic messenger RNA: Functions in initiation and control of translation. Prog Nucleic Acid Res Mol Biol 35: 173–207
Storti RV, Scott MP, Rich A, Pardue ML (1980) Translational control of protein synthesis in response to heat shock in *D. melanogaster* cells. Cell 22: 825–834
Tahara SM, Morgan MA, Shatkin AJ (1981) Two forms of purified m^7G-cap binding protein with different effects on capped mRNA translation in extracts of uninfected and poliovirus-infected HeLa cells. J Biol Chem 256: 7691–7694
Tas PWL, Martini OHW (1987) Regulation of ribosomal protein S6 phosphorylation in heat-shocked HeLa cells. Eur J Biochem 163: 553–559
Thach RE (1992) Cap recap: the involvement of eIF-4F in regulating gene expression. Cell 68: 177–180
Thomas AAM, ter Haar E, Wellink J, Voorma HO (1991) Cowpea mosaic virus middle component RNA contains a sequence that allows internal binding of ribosomes and that requires eukaryotic initiation factor 4F for optimal translation. J Virol 65:2953–2959
Thomas AAM, Scheper GC, Voorma HO (1992) Hypothesis–is eukaryotic initiation factor-2 the scanning factor. New Biol 4: 404–407

Trachsel H, Sonenberg N, Shatkin A, Rose J, Leong K, Bergman J, Gurdon J, Baltimore D (1980) Purification of a factor that restores translation of VSV mRNA in extracts from poliovirus infected HeLa cells. Proc Natl Acad Sci USA 77: 770–776

Tsukiyama-kohara K, Iizuka N, Kohara M, Nomoto A (1992) Internal ribosome entry site within hepatitis C virus RNA. J Virol 66: 1476–1483

Verver J, Le Gall O, van Kammen A, Wellink J (1991) The sequence between nucleotides 161 and 512 of cowpea mosaic virus mRNA is able to support internal initiation of translation in vitro. J Gen Virol 72: 2339–2345

Zapata J, Martinez M, Sierra J (1994) Purification and characterization of eukaryotic polypeptide chain initiation factor 4F from *Drosophila melanogaster* embryos. J Biol Chem 269: 18047–18052

Zapata JM, Maroto FG, Sierra JM (1991) Inactivation of mRNA cap-binding protein complex in *Drosophila melanogaster* embryos under heat shock. J Biol Chem 266: 16007–16014

Zhong T, Arndt KT (1993) The yeast SIS1 protein, a DnaJ homologue, is required for the initiation of translation. Cell 73: 1175–1186

Cap-Independent Translation and Internal Initiation of Translation in Eukaryotic Cellular mRNA Molecules

N. IIZUKA[1], C. CHEN, Q. YANG, G. JOHANNES, and P. SARNOW

1 Introduction

Before RNA transcription is completed by RNA polymerase II, the 5' ends of eukaryotic mRNA molecules are modified. Mediated by a series of enzymatic reactions, a 7-methyl GpppN (in which N can be any nucleotide) "cap" structure is added to the 5' end of each primary transcript (BANERJEE 1980; SHATKIN 1976). This cap structure is thought to protect the mRNA from exonucleolytic degradation (FURUICHI et al. 1977; GREEN et al. 1983; STEVENS 1978) and enhances nuclear processes such as splicing and 3' end processing of pre-mRNAs as well as nucleocytoplasmic transport of mature mRNAs (EDERY and SONENBERG 1985; GEORGIEV et al. 1984; KONARSKA et al. 1984). In the cytoplasm, the cap structure greatly facilitates the binding of ribosomes to mRNAs (BANERJEE 1980; GALLIE

Department of Biochemistry, Biophysics and Genetics, and Department of Microbiology, University of Colorado Health Sciences Center, Denver, CO 80262, USA
[1]Present address: Department of Virology, Medical School, Nagoya City University, Nagoya, Japan

1991; SHATKIN 1976). The cap-binding protein complex eIF-4F, composed of proteins eIF-4E, eIF-4A and eIF-4γ (EDERY et al. 1983; GRIFO et al. 1983; TAHARA et al. 1981), interacts with the cap structures. Because eIF-4γ and eIF-4E have been detected both in the nucleus and in the cytoplasm (ETCHISON and ETCHISON 1987; LEJBKOWICZ et al. 1992), it is thought that the cap binding protein complex interacts with mRNA cap structures in both of these cellular compartments.

The initial finding that picornaviral mRNAs are uncapped (NOMOTO et al. 1976, 1977) showed that certain uncapped mRNAs can be translated in a cap-independent manner in vivo in mammalian cells. Subsequently, it was shown that the cap-independent translation of picornaviral mRNAs proceeded by an internal ribosome binding mechanism (JANG et al. 1988; PELLETIER and SONENBERG 1988; reviewed in the chapters by Hellen and Wimmer, Ehrenfeld and Semler, and Belsham et al.). Since internal initiation of translation could be performed by the cellular translational apparatus in uninfected cells (PELLETIER and SONENBERG 1989), the question arose whether cellular mRNAs could be translated cap-independently as well. This question sounded like an oxymoron, because all eukaryotic mRNAs are, by definition, capped. A low requirement of intact cap-binding protein complex eIF-4F for translation of certain capped mRNAs, notably late adenoviral mRNAs (see chapter by Schneider) and heat shock mRNAs (see chapter by Rhoads and Lamphear), has provided one criterium to describe cap-independent translation of capped mRNAs. It is possible that some capped mRNAs can be translated both by a 5' end-dependent scanning mechanism (KOZAK 1989) and by an internal ribosome binding mechanism like picornaviral RNAs. Such RNAs may contain sequences mediating internal initiation, internal ribosome entry sites (IRESs), in their 5' noncoding regions (5'NCR) that allow them to be translated at times when eIF-4F is nonfunctional and cannot bind to the cap structure (BONNEAU and SONENBERG 1987; HUANG and SCHNEIDER 1991). Alternatively, an IRES located within the coding region of an mRNA could render the mRNA functionally polycistronic, and translation could result in the production of several protein products. This chapter summarizes studies of cap-independent translation and internal initiation of cellular mRNA in mammalian cells and in the budding yeast *Saccharomyces cerevisiae*.

2 Experimental Systems for Studying Cap-Independent Translation and Internal Initiation

Most of the methods used to study cap-independent translation are discussed in the chapter by Jackson and colleagues. We will briefly refer to some of those methods and describe in more detail a novel method, one that uses circular RNA molecules, to study internal initiation.

2.1 Cap Analogs and Hybrid-Arrested Translation

Cap-dependent translational initiation requires the binding of the 40S ribosomal subunit, loaded with translational initiation factors, at the capped 5' end of the mRNA. The recruitment of 40S ribosomal subunits is thought to be facilitated by the interaction with the cap-binding protein complex eIF-4F bound to the terminal cap structure (Hershey 1991; Kozak 1991; Merrick 1992). When cap analogs such as m^{7}GpppG dinucleotides are added to in vitro translation systems, they will compete with the capped mRNAs for binding to the limited supply of eIF-4F, resulting in inefficient binding of 40S subunits to the mRNA and a decrease in translation (Canaani et al. 1976; Weber et al. 1987). Thus, mRNAs that can still be translated efficiently in the presence of cap analogs are thought to require low amounts of eIF-4F. One explanation for a low requirement for eIF-4F is that these mRNAs contain unstructured 5'NCRs that can bind 40S subunits in the absence of significant amount of eIF-4F. The heat shock mRNAs (Maroto Sierra 1988), (see chapter by Rhoads and Lamphear) and the late leaders of adenovirus mRNAs (Dolph et al. 1988, 1990), (see chapter by Schneider) may provide two such examples. Alternatively, a low eIF-4F requirement for translational initiation may indicate that the mRNAs bear sequences that can mediate internal ribosome entry. Examples include: (1) the vesicular stomatitis virus P mRNA (Herman 1987), (2) picornaviral mRNAs (see chapter by Hellen and Wimmer) and (3) certain mRNAs that encode cellular proteins (see below).

The presence of an mRNA-DNA hybrid near the 5' end of an mRNA will result in the inhibition of cap-dependent translation, because the RNA-DNA hybrid will block the scanning 40S subunits from reaching the initiator AUG translational start codon. Thus, failure to inhibit translation by an RNA-DNA hybrid located upstream of the translational start codon is one good indication that the mechanism of translation initiation is cap-independent and possibly occurs by internal initiation (Herman 1986; Sankar et al. 1989; Shih et al. 1987).

2.2 Dicistronic RNAs and Dicistronic Viruses

Many studies have employed dicistronic mRNAs to identify IRES elements that can mediate cap-independent translation (Jang et al. 1988; Pelletier and Sonenberg 1988). The first cistron in a capped dicistronic mRNAs can be translated by a cap-dependent scanning mechanism. The second cistron should not be translated unless preceded by an IRES element. Of course, it is important to test whether the second cistron is indeed translated from the intact dicistronic mRNA and not from uncapped, truncated mRNA species containing the second cistron sequences. One can conclude that second cistron translation is conferred by internal ribosome binding to the IRES element if only intact dicistronic mRNAs are associated with polysomes under conditions in which cap-dependent translation is inhibited, for example, in poliovirus -infected cells (Macejak and Sarnow 1991; Oh et al. 1992; Pelletier and Sonenberg1988).

The presence of functional IRES elements can be verified by inserting the putative IRES into the plus-stranded poliovirus RNA genome (MOLLA et al. 1992, 1993). Normally, the viral mRNA encodes a 220 kDa polyprotein whose NH_2-terminal portion encodes the structural proteins of the viral particle. The COOH-terminal portion of the polyprotein encodes the replication proteins and the proteases that are involved in the processing of the polyprotein. Molla and coworkers disrupted the polyprotein coding region by insertion of an IRES element between the capsid protein-coding and replication protein-coding open reading frame. This artificial dicistronic RNA could be replicated after introduction into mammalian cells and subsequently packaged into virions. Amplification of the dicistronic RNAs was only possible if the introduced IRES mediated translation of the viral replication proteins. That intact, and not broken, dicistronic mRNAs were responsible for the initiation of second cistron translation was suggested by the fact that the number of plaque-forming viruses was linearly dependent on the virus stock concentration, indicating that each individual plaque was the result of infection by a single poliovirus particle carrying a single dicistronic RNA molecule (MOLLA et al. 1992). Although it is possible that subgenomic mRNAs were generated during the infectious cycle that could have functioned as template RNAs for translation of nonstructural proteins, the initial protein products clearly had to be translated from the incoming full-length dicistronic mRNA.

2.3 Circular RNAs

As discussed above, translational studies with dicistronic mRNAs can reveal the presence of sequence elements that mediate internal initiation. However, dicistronic mRNAs are of only limited use in studying the mechanism of ribosomal positioning by the IRES at the AUG start codon of the second cistron. This is due to the possibility that 40S subunits, instead of binding directly to the IRES, could first bind to the 5' end of the capped or uncapped mRNA and subsequently be transferred to the IRES. Monitoring the translation of IRES-containing mRNAs whose 5' ends were blocked in some way that prevented interaction with 40S subunits would help in distinguishing between these possibilities.

It has been known for some time that eukaryotic ribosomes are unable to bind to small circular RNAs, 25–110 nucleotides in length, in a cell-free system; in contrast, prokaryotic ribosomes can bind to such small circular RNAs, although with low efficiency (KONARSKA et al. 1981; KOZAK 1979). From this, it was concluded that eukaryotic ribosomes could only bind to RNA via free 5' ends. However, one would predict that IRES elements could function in RNAs without free 5' ends.

To test this prediction, we developed a method to produce large quantities of circular RNAs up to 1000 nucleotides in length (CHEN et al. 1993; CHEN and SARNOW 1995). We used a modification of a method originally described by MOORE and SHARP (1992), outlined in Fig. 1. Briefly, DNA plasmids containing the promoter for T7 RNA polymerase were linearized and used as templates for in vitro transcription

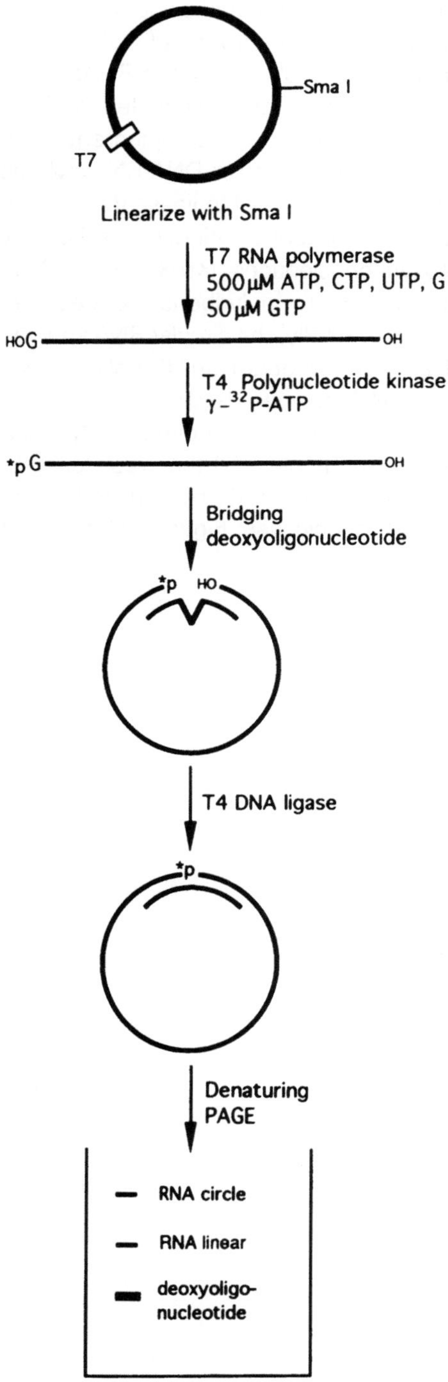

Fig. 1. Modified Moore and Sharp method (Moore and Sharp 1992) for the production of circular RNAs. (From Chen et al. 1993)

reactions containing 500 μ*M* each of CTP, ATP, UTP and guanosine nucleoside, but only 50 μ*M* of GTP. Thus, most of the transcripts begin with a 5' OH. Subsequently, radiolabeled phosphates were added to the 5' ends of the RNAs by T4 polynucleotide kinase. The radiolabeled RNAs were then annealed to DNA "splints," deoxyoligonucleotides that contained sequences complementary to both the 5' and 3' ends of the RNAs. The resulting circular DNA-RNA hybrids (Fig. 1) were then covalently closed by the addition of T4 DNA ligase (MOORE and SHARP 1992), which has been shown to ligate oligoribonucleotides efficiently when hybridized to complementary deoxyribonucleotides (KLEPPE et al. 1970). Interestingly, T4 DNA ligase has a much lower K_m for polyribonucleotides than does RNA ligase (ROMANIUK and UHLENBECK 1983). Finally, the circular RNA species were separated in urea-containing polyacrylamide gels from the oligodeoxynucleotides and unligated or broken linear RNA species. Radiolabeled circular RNA species were distinguished from linear monomeric and dimeric RNAs by: (1) their migration in urea-containing polyacrylamide gels, (2) their resistance to dephosphorylation by bacterial alkaline phosphatase and (3) by their RNA fragmentation pattern after digestion with endoribonucleases. Figure 2 shows that

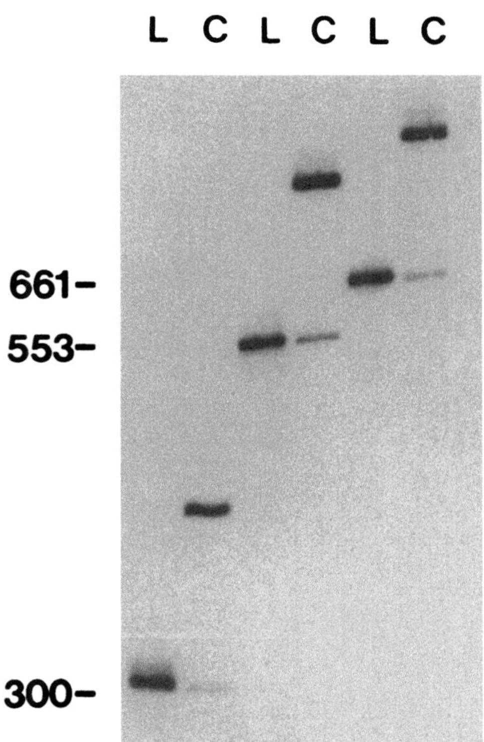

Fig. 2. Analysis of circular RNAs in urea-containing polyacrylamide gels. Production of RNA circles 300, 553 and 661 nucleotides in length, in pairs of lanes from *left* to *right*. An autoradiograph displaying the linear (L) and circular (C) RNA specifies is shown. (from CHEN et al. 1993)

this method is useful for the production of RNA circles ranging from 300 to 661 nucleotides in length. Circles up to 1200 nucleotides in length can be obtained with lower yields (Chen and Sarnow, unpublished).

We have constructed circular RNAs containing the IRES element present in the 5'NCR of the encephalomyocarditis virus RNA genome (JANG et al. 1988; JANG and WIMMER 1990) and incubated these circles with eukaryotic ribosomes in the presence of the elongation inhibitor sparsomycin. When the fate of the circles in translation extracts was monitored by sedimentation in sucrose gradients, IRES-containing circles were detected in a fraction sedimenting at 80S, indicating that an intact ribosome was attached. After incubation in a rabbit reticulocyte translation lysate, the IRES-containing circles mediated translation of the predicted protein product (CHEN and SARNOW 1994). Furthermore, we have monitored the translation of circular IRES-containing RNAs containing a continuous open reading frame. Analyses of the translation products demonstrated that ribosomes were able to translate the circles many times resulting in the production of a very long polypeptide chain (CHEN and SARNOW 1995). Thus, IRES elements can mediate translational initiation in the absence of a free 5' end in the mRNA, probably by direct recruitment of ribosomal subunits. These findings indicate that RNA circles should aid in studying the mechanism with which IRES elements mediate translation initiation and should provide valuable tools for identifying further RNA sequences that can function as IRES elements.

2.4 Poliovirus-Infected Cells

Since intact cap binding protein complex eIF-4F is present only in limited amounts in infected cells due to virus-induced proteolysis of the p220 component of eIF-4F (ETCHISON et al. 1982), eukaryotic mRNAs that have a reduced requirement for intact eIF-4F are preferentially translated in poliovirus-infected cells. Indeed, several mRNAs with reduced secondary structures in their 5' NCRs have been found to be translated in poliovirius-infected cells at a time when p220 is proteolyzed. Such mRNAs include heat shock protein 70 mRNAs (MUNOZ et al. 1984) and late adenovirus mRNA containing the tripartite leader (DOLPH et al. 1988, 1990).

In addition, mRNAs that contain, like poliovirus RNA, an IRES element should be translated in virus-infected cells at a time when the bulk of eIF-4F is not intact. Indeed, the first cellular mRNA containing an IRES was identified by its translation in poliovirus-infected cells (SARNOW 1989).

3 Identification of Cellular Internal Ribosome Entry Site Elements

In studying the phenotypes of several poliovirus mutants which were defective in viral RNA amplification, it was noted that the production of a 78 kDa cellular protein was not inhibited in mutant-infected cells when the translation of most cellular mRNAs was inhibited (SARNOW 1989). Subsequently, the 78 kDa protein was identified as glucose-regulated protein 78 (LEE et al. 1981, 1984), also known as immunoglobulin heavy chain binding protein (BiP) (MUNRO and PELHAM 1986). It was found that they steady state level of BiP mRNA was the same in infected and uninfected cells; nevertheless, translation of the BiP mRNA was enhanced in mutant-infected cells compared to uninfected cells (SARNOW 1989). This suggested that the BiP mRNA could be translated cap-independently like poliovirus mRNA. Subsequently, it was tested whether the mechanisms of translational initiation of BiP and poliviral mRNA were similar.

3.1 The BiP Internal Ribosome Entry Site

Inspection of the human BiP gene sequence revealed a 220 nucleotide 5'NCR devoid of AUG codons upstream of the initiator AUG codon (TING and LEE 1988). To examine the mechanism of translational initiation of BiP mRNA, RNA molecules that contained the BiP 5'NCR upstream of a luciferase coding region were synthesized in vitro by T7 RNA polymerase. These RNAs were then transfected into mammalian cells that had previously been infected with poliovirus. It was found that RNAs containing the BiP 5'NCR were translated in infected cells, while RNAs bearing different 5'NCRs were not translated (MACEJAK et al. 1990). These results showed that the 5'NCR of BiP was sufficient to mediate translation in poliovirus-infected cells, even though the p220 component of eIF-4F was proteolyzed and translation of most cellular mRNAs was inhibited.

Next, it was tested whether the observed eIF-4F-independent translation mediated by the BiP 5'NCR was accomplished by an internal initiation mechanism. When the translation of dicistronic mRNAs containing the BiP 5'NCR in the intercistronic spacer region was monitored in mammalian cells, it was found that the second cistron was translated when preceded by either the BiP 5'NCR or the poliovirus 5'NCR. However, no second cistron translation was observed when the intercistronic spacer region contained any of several other unrelated RNA sequence elements of similar or shorter lengths as the BiP 5'NCR (MACEJAK and SARNOW 1991). The introduction of an RNA hairpin structure at the very 5' end of the first cistron of the dicistronic RNA containing the BiP 5'NCR completely abolished the translation of the first cistron but had no effect on the efficiency with which the second cistron was translated (MACEJAK and SARNOW 1991). This suggested that the BiP 5'NCR does not facilitate readthrough of ribosomes from the first to the second cistron; instead, the BiP 5'NCR can function as an IRES element.

Current research focuses on the identification of protein factors that interact with the BiP IRES. We noted that the BiP IRES functioned poorly when the RNAs were introduced directly into the cytoplasm. In contrast, BiP IRES-containing RNAs expressed in the nucleus as polymerase II transcripts were translated very well after their translocation into the cytoplasm (Macejak and Sarnow, unpublished). These findings suggested that BiP IRES-containing RNAs may recruit nuclear factors that are subsequently involved in cytoplasmic translational initiation. Two nuclear factors, the La antigen (MEEROVITCH et al. 1993) and the PTB polypyrimidine binding protein (HELLEN et al. 1993), have been found to bind to picornavirus IRES elements (see chapters by Hellen and Wimmer and by Belsham et al.); PTB, however, binds extremely poorly to the BiP IRES (E. Wimmer, personal communication) and, therefore, is not a likely candidate to modulate internal initiation mediated by the BiP IRES. We have continued to search for nuclear proteins from human HeLa cells that can specifically interact with the BiP IRES. Two proteins that can bind specifically to the BiP IRES, approximately, 60 kDa and 95 kDa in size, have been detected in partially purified extracts. It is not yet clear whether the specific binding of p95 and p60 to the BiP IRES is involved in IRES function. Interestingly, both p95 and p60 could bind to two independent domains in the BiP IRES (Fig. 3). Both RNA fragments spanning nucleotides 1–127 and 128–220 could be cross-linked to these proteins. Because each of these RNA domains could direct internal initiation when placed into the intercistronic spacer region of a dicistronic mRNA, a correlation between protein binding and internal initiation sites can be made. That both p95 and p60 can also bind to IRES elements present in poliovirus (PELLETIER and SONENBERG 1988), encephalomyocarditis virus (JANG et al. 1988) and hepatitis C virus (TSUKIYAMA-KOHARA et al. 1992; WANG et al. 1993) makes p95 and p60 good candidates for modulating internal initiation (Yang and Sarnow, unpublished).

3.2 The *Antennapedia* Internal Ribosome Entry Site

Searching for further candidate mRNAs that contain IRES elements, it was brought to our attention that many *Drosophila* mRNAs contain long 5'NCRs often burdened with several AUG codons. Cavener complied the features of 403 different mRNAs noting that the average *Drosophila* 5'NCR contains 248 nucleotides and that 42% contain one or more AUG codons upstream of the predicted start codon (CAVENER and CAVENER 1993). In contrast, only 9% of characterized vertebrate mRNAs contain long leader sequences with upstream AUG codons (KOZAK 1987a). Most certainly, there is a bias in the *Drosophila* database collection towards genes involved in developmental regulation, because most *Drosophila* genes were identified following the characterization of mutant phenotypes.

The *Antennapedia (Antp)* gene of *Drosophila melanogaster* provides an example of mRNAs with exceptionally long 5'NCRs. Two promoters produce transcripts with quite different 5'NCRs and identical coding regions (LAUGHON et al. 1986; STROEHER et al. 1986). Promoter P1-derived mRNAs are 1512 nucleotides in length and comprise upstream noncoding exons A, B, D and part of E. Promoter P2-derived mRNAs, 1727 nucleotides in length, contain noncoding exons C,D and part of E. To translate P2-derived mRNAs by the scanning mechanism (KOZAK 1989), 43S ternary complexes would have to scan 1727 nucleotides and bypass 15 AUG codons, six of which are embedded in consensus sequences for translational initiation in *Drosophila* (CAVENER 1987). A similarly difficult scenario exists for

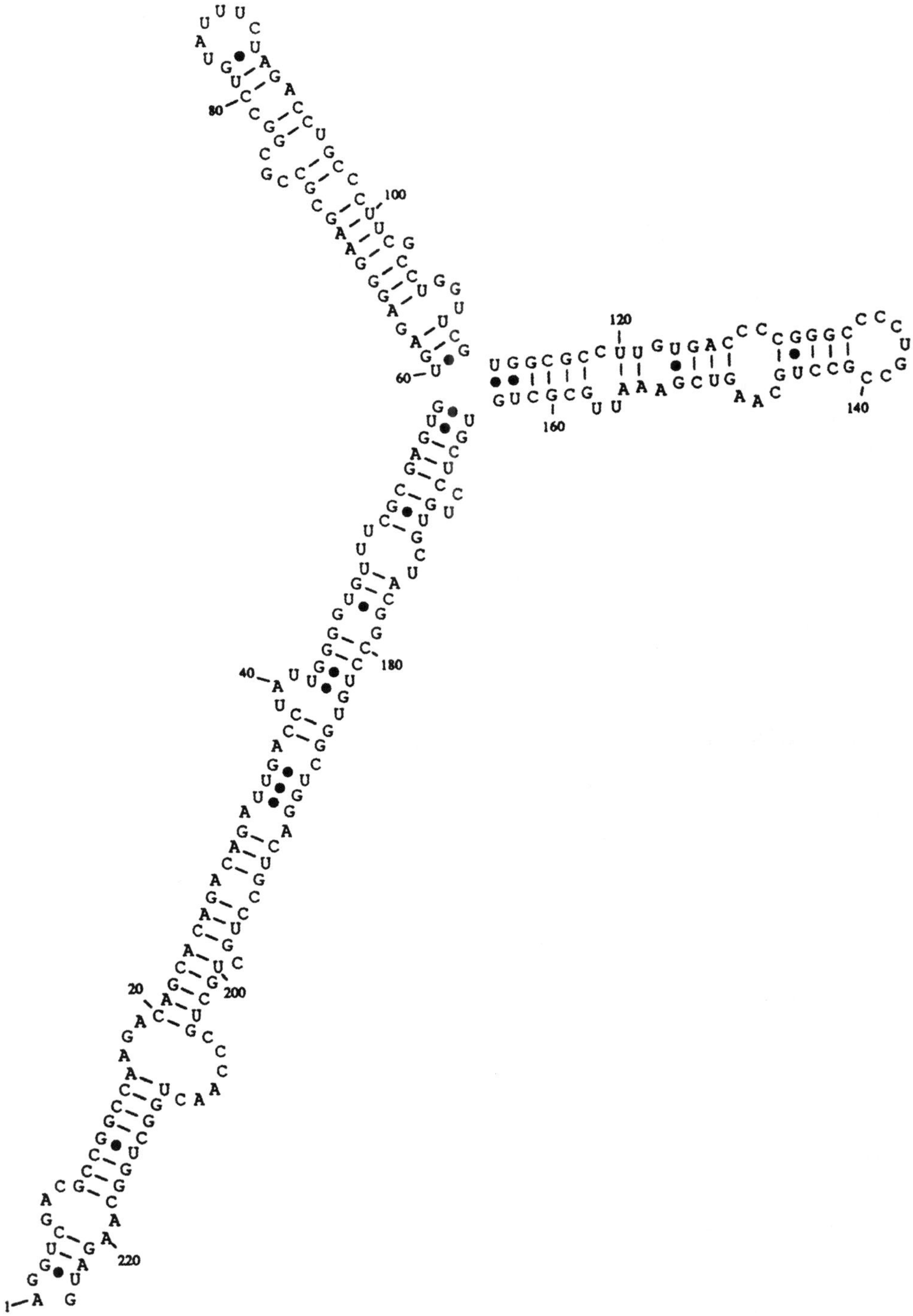

Fig. 3. Sequence and predicted structure (ZUKER and STIEGLER 1981) of the BiP internal ribosome entry site. The 3'-terminal nucleotides represent the AUG translational start codon. This structure has a predicted lowest free-energy value of $\Delta G = -72.4$ kcal/mol

P1-derived mRNAs. Thus, *Antp* mRNAs are predicted to be translated extremely poorly by a scanning mechanism (KOZAK 1989). Alternatively, Antp mRNAs could be translated efficiently either if the 5'NCRs harbored IRES elements or if they contained features that allowed scanning 43S complexes to bypass long structural elements, known as a "shunting" mechanism (FÜTTERER et al. 1993).

To test whether *Antp* mRNAs contain an IRES element, the translation of dicistronic mRNAs containing the 5'NCR of the P2-derived transcript in the intercistronic spacer region was monitored in cultured *Drosophila* cells. It was found that the sequences encoded by exons D and E (Fig. 4), present in both P1- and P2-derived transcripts, could function as an IRES (OH et al. 1992). Insertion of an AUG codon into the 3' border of the 250 nucleotide exon D-E sequence resulted in the usage of this AUG as a translational start codon. This indicated that the 5' border of exon D functions as the entry site for 43S ribosomal subunits. Inspection of sequences in exon D-E revealed the presence of a 55 nucleotide RNA sequence that is highly conserved (nucleotides 41–96 in Fig. 4) among *Drosophila* species whose common ancestors lived 60 million years ago (HOOPER et al. 1992). Specifically, there are only four nucleotide changes in the 55 nucleotide RNA sequence from *Drosophila melanogaster, Virilis* and *subobscura* (HOOPER et al. 1992). Preliminary experiments have shown that this 55 nucleotide RNA sequence is required for *Antp* IRES function (Oh and Sarnow, unpublished).

It is possible that translational initiation by internal ribosome binding in *Drosophila* is used at particular times during development. For example, many maternal and early zygotic mRNAs are thought to be translated when the embryo is at a single cell syncytium undergoing rapid mitotic divisions. However, in mammalian cells, cap-dependent translation is known to be severely impaired during mitosis, because of the underphosphorylation of the capbinding protein eIF-4E (BONNEAU and SONENBERG 1987; HUANG and SCHNEIDER 1991). If the same is true in the *Drosophila* embryo, it is not clear how early embryonic mRNAs such as *biocoid* (BERLETH et al. 1988) and *nanos* (WANG and LEHMANN 1991) are translated efficiently. A cap-independent translation mechanism such as internal ribosome binding could provide a means by which mRNAs are translated in the absence of functional, cap binding protein complexes.

4 Translation in *Saccharomyces cerevisiae*

The major features of the translational apparatus in yeast and in multicellular eukaryotes are similar (CIGAN and DONAHUE 1987). In fact, homologs for many genes that encode translation factors in higher eukaryotes have been found in yeast (Table 1), suggesting that the initiation, elongation and termination pathways of protein biosynthesis are quite similar.

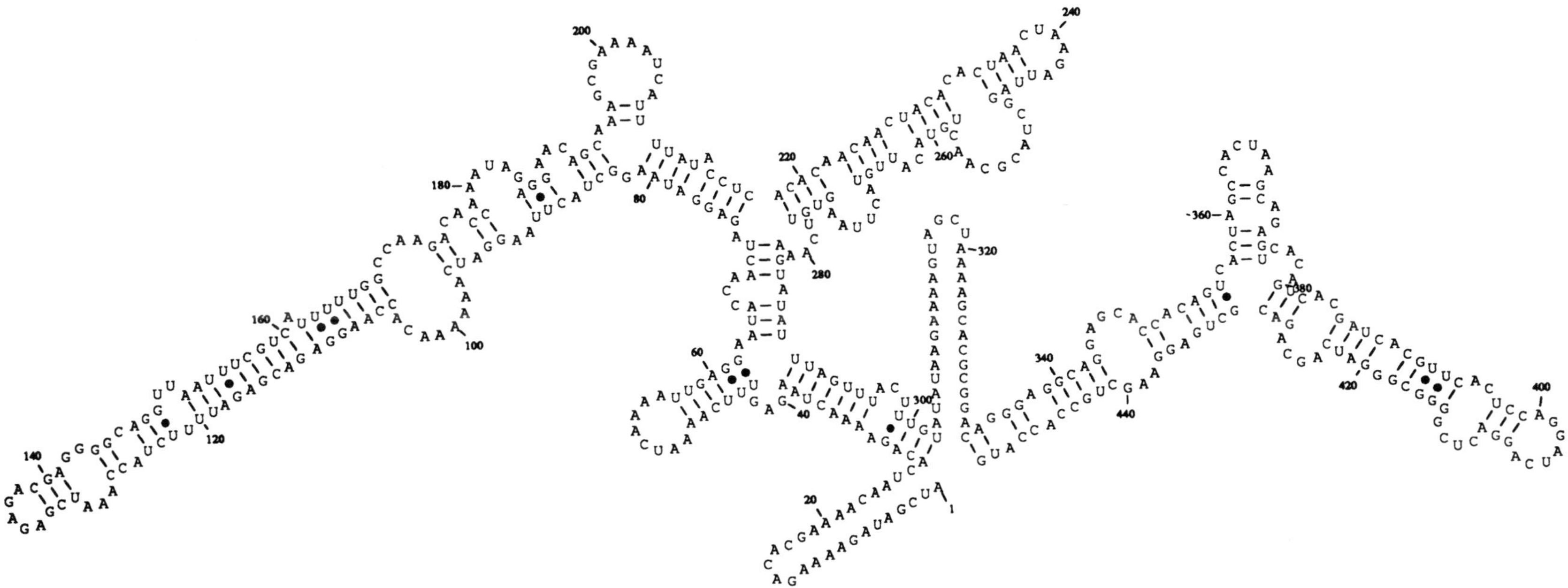

Fig. 4. Sequence and predicted structure (ZUKER and STIEGLER 1981) of the *Antennapedia* internal ribosome entry site. Exon D starts with nucleotide numbered 41. The 3'-terminal nucleotides represent the AUG translational start codon. This structure has a predicted lowest free-energy value of $\Delta G = -70.2$ kcal/mol

Table 1. Translational initiation factors in *Saccharomyces cerevisiae*

Yeast gene	Vertebrate homolog	Properties/method of isolation	Reference
SUI2 (l)	eIF2-α	Suppressor of initiation codon mutation	CIGAN et al. (1989)
SUI3 (l)	eIF2-β	Suppressor of initiation codon mutation	DONAHUE et al. (1988)
GCD11(l)	eIF2-γ	Negative regulator of *GCN4* expression	HANNIG et al. (1993)
GCD1(l), *GCD2*(l) *GCD6*(l), *GCD7* (l) *GCN3* (ne)	eIF-2B	Regulators for *GCN4* expression Yeast guanine nucleotide-exchange factor	CIGAN et al. (1993)
TIF1 (ne), *TIF2* (ne) (*tif1, tif2:* l)	eIF-4A	Suppressors of mitochondrial missense mutations	LINDER and SLONIMSKI (1989)
TIF3 (sg, cs)	eIF-4B	Isolated with an antibody directed against eIF-4E	ALTMAN et al. (1993)
STM1 (sg, cs)	eIF-4B	Isolated as suppressor of eIF-4A	COPPOLECHIA et al. (1993)
eIF-4E(l)	eIF-4E	Isolated by screening of a yeast cDNA library in λgt11 with affinity purified anti-eIF-4E antibody	ALTMANN et al. (1987)
CDC33	eIF-4E	Isolated as cell cycle mutant	BRENNER et al. (1988)
TIF4631 (sg, cs) *TIF4632* (ne) (*tif4631, tif4632*:l)	eIF-4Fγ	Isolated with an antibody directed against eIF-4E	GOYER et al. (1993)
TIF5(l)	eIF-5	Catalyzes 80S formation	CHAKRAVARTI and MAITRA (1993)
TIF1A(sg) *TIF51B*(ne) (*tif51A, tif51B:* l)	eIF-5A	Its hypusine modification is essential for cell viability	SCHNIER et al. (1991)
SUI1 (l)	?	Suppressor of initiation codon mutation	YOON and DONAHUE (1992)
SSL1 (l)	?	Suppressor of RNA hairpin structure in 5' noncoding region; affects UV resistance	YOON et al. (1992)
SSL2 (l)	ERCC-3	Suppressor of RNA hairpin structure in 5' noncoding region	GULYAS and DONAHUE (1992)
GCN2(ne)	DAI-kinase	Phosphorylates eIF2-α	DEVER et al. (1992)
GCN1 (ne)	homolog of fungi EF-3	Required for GCN2 activation	MARTON et al. (1993)
SIS1(l)	homolog of DnaJ	Required for translational initiation	ZHONG and ARNDT (1993)
PRT1(l) = *CDC63*	?	Affect translational initiation	HANIC-JOYCE et al. (1987)

Phenotypes of mutants: l, lethal; sg, slow growth; cs, cold-sensitive; ne, no effect.

4.1 Comparison of 5'NCRs in Lower and Higher Eukaryotes

Yeast mRNAs have some properties that are distinct from mRNAs in higher eukaryotes. Specifically, the chemical composition of the 5'-terminal cap structure, the length and nucleotide composition of the 5'NCR and the nucleotide sequences surrounding the translational start AUG codon vary between yeast and higher eukaryotic mRNAs.

The cap structure, present at the 5'end of mRNAs, is known to effect translation initiation by enhancing the rate of initiation and conferring stability of mRNAs (EVERETT and GALLIE 1992; GALLIE 1991). The cap structures on higher eukaryotic mRNAs are usually composed of $m^7GpppN_1mN_2$ or $m^7GpppN_1m\text{-}N_2m$ 5'-terminal nucleotides (in which N can be any nucleotide and m denotes a methyl group). In contrast, yeast mRNAs contain more simple m^7GpppN cap structures, known as cap-0 structures (DE KLOET and ANDREAN 1976; SRIPATI et al. 1976).

At less than 100 nucleotides, the average length of a yeast 5'NCR is slightly shorter than that of a higher eukaryotic mRNA (CIGAN and DONAHUE 1987). Short 5'NCRs (<21 nt) have been shown to confer reduced translation to mRNAs in yeast, probably due to the fact that 80S ribosomes sterically inhibit the binding of 43S complexes at the 5' end of the mRNA (VAN DEN HEUVEL et al. 1989). However, there is also precedence that yeast *tcm 1* mRNA can be translated in the complete absence of a 5'NCR (MAICAS et al. 1990).

Yeast 5'NCRs are enriched in adenine residues, suggesting that they are relatively devoid of significant secondary structures. In fact, the predicted average free energy for a yeast 5'NCR is –7 kcal/mol. Insertions of stable RNA hairpin structures into the 5'NCRs of *CYC1* and *HIS4* reduced their translational efficiency by 20- to 100-fold (BAIM and SHERMAN 1988; CIGAN et al. 1988b). The presence of an RNA hairpin in a yeast 5'NCR with a predicted free energy of –28kcal/mol inhibited translation by at least 98% (VEGA LASO et al. 1993). These effects are likely to be due to the inhibition of scanning of 43S subunits, supporting the conclusion that most yeast mRNAs are translationally initiated by the scanning mechanism as are most mRNAs in higher eukaryotes (KOZAK 1989).

In contrast to the vertebrate sequence ACCA*AUG*G (KOZAK 1986, 1987) surrounding the translational start AUG codon, the yeast translational initiation start site consensus is (A/Y)A(A/U)A*AUG*UCU (CIGAN and DONAHUE 1987). In addition, the effect of altered context sites on translational efficiency is less pronounced in yeast than in higher eukaryotes (BAIM and SHERMAN, 1988; CIGAN et al. 1988; SHERMAN and STEWART 1982). The presence of upstream AUG codons in the 5'NCRs is rare; only approximately 5% of yeast genes contain upstream open reading frames, compared with 9% of characterized vertebrate mRNAs (KOZAK 1987) and 42% of *Drosophila* mRNAs (CAVENER and CAVENER 1993).

4.2 Genes Involved in Translational Initiation

Using genetic approaches, many yeast genes encoding translational initiation factors have been identified. Table 1 lists some of these initiation factors. For a more detailed description of some of these factors, the reader is refered to reviews by CIGAN and DONAHUE (1987) and HINNEBUSCH 1990 (HINNEBUSCH and LIEBMAN 1991).

Two sets of genetic approaches have been crucial to the identification of the yeast homologs of the eIF-2 complex and its regulators. First, a selection was employed for suppressors that restored the translation of a *HIS4* yeast gene that lacked an AUG initiator codon (CIGAN et al. 1988b). In this way, *SUI2* (eIF-2α) (CIGAN et al. 1989), *SUI3* (eIF-2β) (DONAHUE et al. 1988), and *SUI1*, an additional factor that functions in concert with eIF-2 (YOON and DONAHUE 1992), were identified. These experiments clearly showed that other factors besides $tRNA_i^{Met}$ (CIGAN et al. 1988a) are involved in directing the ribosome to the correct translational start site for protein biosynthesis. In another approach, the isolation and characterization of both positive and negative regulators of *GCN4* expression have uncovered many translational initiation factors such as eIF-2γ (HANNIG et al. 1993), each member of the eIF-2B complex (CIGAN et al. 1993), the eIF-2 kinase, *GCN2*, (DEVER et al. 1992) and its activator *GCN1* (MARTON et al. 1993).

TIF1 and *TIF2*, both homologs of eIF-4A, were identified as suppressors of missense mutations in the mitochondrial *oxi2* gene (LINDER and SLONIMSKI 1989). Interestingly, *STM1* (eIF-4B) was isolated as a suppressor of a temperature-sensitive *tif1* mutant (COPPOLECCHIA et al. 1993), possibly suggesting a physical interaction between the two gene products.

The genes encoding polyadenosine binding protein (PAB; SACHS and DAVIS 1989) and its associated nuclease. PAN; SACHS and DEARDORFF 1992) are not listed in Table 1. Genetic analysis has shown that although both of these proteins play important roles in translational initiation, their exact function(s) remains unknown, and no homologs have been identified in higher eukaryotes.

4.3 Mechanisms of Initiation

4.3.1 Cap-Dependent Initiation

Most yeast mRNAs are likely to be translated by the cap-dependent, 5' end-dependent scanning mechanism known to operate in the translation of most higher eukaryotic mRNAs (CIGAN and DONAHUE 1987; HINNEBUSCH and LIEBMAN 1991; KOZAK 1989). As in higher eukaryotes, binding of a cap binding protein complex to the capped mRNAs is a prerequisite for recruitment of 40S ribosomal subunits. In contrast to the higher eukaryotic cap binding protein complex eIF-4F, composed of eIF-4E, eIF-4A and eIF-4γ (p220), yeast eIF-4F contains eIF-4B (TIF3) instead of eIF-4A (ALTMANN et al. 1993). From this, it was concluded that yeast eIF-4B contributes to the known RNA helicase activity of eIF-4F (ROZEN et al. 1990) thought to be important for the unwinding of secondary structure present in the 5'NCRs of the mRNAs. The 40S ribosomal subunits, recruited onto the mRNAs near their 5' ends, subsequently scan the mRNA in a 5' to 3' direction until an appropriate AUG is encountered which is used as the start site of protein synthesis.

4.3.2 Initiation by Reinitiation

The yeast GCN4 mRNA contains a 5'NCR that is 590 nucleotides in length. This unusually long leader contains four open reading frames. The synthesis of the GCN4 protein, initiated at the fifth open reading frame, occurs only during amino acid starvation (HINNEBUSCH 1990).

Over the past few years, Hinnebusch and coworkers have elucidated the mechanism of GCN4 mRNA translation and its regulation. Briefly, under nonstarvation conditions, 40S subunits are thought to scan the GCN4 5'NCR in a 5' to 3' direction. After translation of the first open reading frame, 40S subunits resume scanning of the 5'NCR, reacquire initiator tRNA and initiation factors and are competent to reinitiate protein synthesis at the fourth open reading frame, after which 40S dissociate from the mRNA. Thus, the fifth open reading frame, that of the GCN4 protein, is not translated. During amino acid starvation conditions, reinitiation does not occur until the fifth open reading is reached by the scanning 40S. This delay is caused by the accumulation of uncharged tRNA molecules, which result in the activation of GCN2 kinase which phosphorylates eIF-2α. Phosphorylated forms of eIF-2α sequester eIF-2B, a factor that is involved in recycling eIF2-GDP to eIF2-GTP. As a consequence, only low concentrations of eIF2-GTP complexes are present in amino acid-deprived cells, reloading of "empty" scanning 40S complexes is delayed, and reinitiation occurs at the fifth instead of the fourth open reading frame of the GCN4 mRNA (DEVER et al. 1992).

4.3.3 Translation of Double-Stranded RNAs of Virus-Like Particles

Most laboratory strains of *S. cerevisiae* have cytoplasmic virus-like particles (VLPs) whose genomes consist of double-stranded RNAs (BRUENN 1980; WICKNER 1992). The best characterized species of VLPs are known as L-A and M1. L-A particles contain a single 4600 base pair double-stranded RNA, M1 VLPs contain one or two double-stranded RNAs, 1800 base pairs in length. M1 RNA encodes a secreted toxin, responsible for the "killer" phenotype of M1-containing cells. M1 is a satellite of L-A and is encapsidated in particles composed of the L-A -encoded coat proteins (BRUENN 1980; WICKNER 1992).

The genomic RNAs of L-A and M are naturally uncapped (BRUENN and KEITZ 1976). Thus, it seems likely that L-A and M RNAs are translated cap-independently, although it has not been tested whether polysomal RNAs are uncapped as well. Russell and coworkers (RUSSELL et al. 1991) have introduced capped and uncapped luciferase mRNAs bearing L-A and M1 5'NCRs into yeast spheroplasts and subsequently monitored the translational efficiencies of these RNA species. It was found that the 5'NCRs of L-A and M1 did not confer a special cap independence to these hybrid mRNAs; however, the poliovirus 5'NCR did not confer cap independence in this transient expression system either (RUSSELL et al. 1991). The latter finding may have been due to the presence of an RNA species that is known to inhibit the translation of RNAs containing the poliovirus IRES specifically (COWARD and DASGUPTA 1992).

That L-A mRNA could be translated cap-independently in vivo has been suggested by studies of host genes that are involved in the maintenance of VLPs. One of these genes is known as *SK12* (WIDNER and WICKNER 1993). This gene was identified because mutant alleles confer a super-killer phenotype to yeast, in which the copy number of L-A and M RNAs is greatly increased. To test the effect of *ski* mutants on translation of capped and uncapped RNAs, RNAs were expressed under the control of polymerase I or polymerase II promoters. Translation of RNA polymerase I-derived, uncapped viral transcripts, but not that of RNA polymerase II-derived, capped viral mRNAs, occurred in *ski2* cells, suggesting that *SKI2* normally represses the translation of uncapped mRNAs (WIDNER and WICKNER 1993). For example, SK12 may hydrolyze uncapped viral RNAs like the gene product of XRN1, a 5' to 3' exonuclease which is responsible for the degradation of uncapped RNAs in yeast (HSU and STEVENS 1993; MUHLRAD et al. 1994; STEVENS and MAUPIN 1987).

4.3.4 Cap-Independent Initiation

Cap-independent translation in yeast extracts mediated by several different viral 5'NCRs has been reported. For example, ALTMANN and colleagues (1989, 1990) prepared translation-competent extracts from yeast cells that were depleted of eIF-4E. In these extracts, cap-independent translation of alfalfa mosaic RNA4 and of mRNAs containing the 5'NCR or poliovirus was very efficient, while the translation of several non-yeast reporter mRNAs was cap-dependent (ALTMANN et al. 1989, 1990). In contrast, COWARD and DASGUPTA (1992) did not detect any translation of polioviral mRNAs both in vivo in yeast cells or in vitro in yeast extracts. The investigators attributed the inhibition of polioviral mRNA translation to the presence of a small RNA inhibitor, which was subsequently characterized. Curiously, translation mediated by the encephalomyocarditis IRES was not inhibited in these extracts (COWARD and DASGUPTA 1992).

Recently, we have established an in vitro translation system from yeast extracts and tested the translation of capped and uncapped mRNAs lacking or containing IRES elements (IIZUKA et al. 1994). Several capped reporter mRNAs could be translated in this extract at least fivefold better than uncapped mRNAs and cap-dependent translation was abolished when 1 m*M* of the cap analog m^7GpppG was added to the system. We also observed efficient translation in this system of mRNAs containing the IRES elements of hepatitis C virus and coxsackievirus. These IRES elements mediated translation when located at the very 5' end of a reporter message or when placed between two open reading frames in dicistronic mRNAs. This finding strongly suggests that the yeast translational apparatus is able to perform cap-independent translation by internal ribosome binding. Because the IRES elements of coxsackievirus and of poliovirus are very similar in sequence and structure, we expected that the presence of the poliovirus IRES inhibitor, noted by COWARD and DASGUPTA (1992), would inhibit the function of the coxsackievirus-IRES as well. However, it is possible that different yeast strains contain different amounts of the poliovirus-IRES inhibitor, that this inhibitor was lost or inactivated during preparation of the extracts, or that the inhibitor is extremely specific for polioviral RNA.

We have searched for naturally occurring yeast 5'NCRs that might mediate cap-independent translation. Using the in vitro translation system as an assay, we showed that the 5'NCRs of two yeast genes. *TFIID* and *HAP4*, supported translation at an enhanced rate in the presence of cap analogs, conditions under which the translation of reporter RNAs lacking these noncoding sequences and the majority of polydenylated yeast mRNAs were inhibited. This cap-independent translation was probably due to internal initiation, because these 5' NCRs also stimulated the translation of the second cistron in dicistronic mRNAs. However, the mechanism of cap-independent translation of mRNAs bearing *TFIID* and *HAP4* 5'NCRs is not known yet; clearly, both noncoding regions do not require significant amounts of cap binding protein complexes to mediate translational initiation. We are currently studying the physiological role of cap-independent translation in yeast, focussing on *TFIID* and *HAP4* mRNAs as promising candidates.

5 Summary and Future Directions

Ample evidence has accumulated that capped eukaryotic mRNAs can be translated without significant amounts of intact cap binding protein complex eIF-4F. Some of these mRNAs are translated by a 5' end-dependent scanning mechanism, while other mRNAs use an internal initiation mechanism reminiscent of prokaryotic translational initiation. An intense search is being undertaken for factors that bind to IRES elements and mediate internal initiation. This is a formidable task that requires elaborate purification schemes and faithful in vitro translation assays. The use of a genetically manipulatable system such as yeast may help in identifying genes whose products are involved in internal initiation. Identification of such yeast genes and isolation of their higher eukaryotic homologs may help in understanding the molecular mechanism of internal initiation. Another important question addresses whether the cap-independent translation and internal initiation mechanisms are regulated during cell growth. Again, genetically manipulatable systems such as yeast and *Drosophila* may provide useful answers. That cap-independent translation is greatly diminished during mitosis in mammalian cells may be starting point to investigate whether cap-independent translation is predominately a mitotic event or whether other physiological roles for cap-independent translation exist.

Acknowledgments. We would like to thank Karla Kirkegaard for many stimulating discussions, and comments on the manuscript. We are also grateful to Keith Gulyas for critical reading of the manuscript. Work described in the authors laboratory was supported by grants from The Council for Tobacco Research USA and from the US Public Health Service. N.I. was supported by a long-term fellowship of the Human Frontier Science Program Organization. G.J. was supported by a NRSA award from the NIH. P.S. acknowledges the receipt of a Faculty Research Award from the American Cancer Society.

References

Altmann M, Sonenberg N, Trachsel H (1989) Translation in Saccharomyces cerevisiae: initiation factor 4E-dependent cell-free system. Mol Cell Biol 9: 4467–4472

Altmann MS, Blum J, Pelletier N, Wilson TMA, Trachsel H (1990) Translation initiation factor-dependent extracts from Saccharomyces cerevisiae. Biochim Biophys Acta 1050: 155–159

Altmann MP, Muller PP, Wittmer B, Ruchti F, Lanker S, Trachsel H (1993) A Saccharomyces cerevisiae homologue of mammalian translation initiation factor 4B contributes to RNA helicase activity. EMBO J 12: 3997–4003

Baim SB, Sherman F (1988) mRNA structures influencing translation in the yeast Saccharomyces cerevisiae. Mol Cell Biol 8: 1591–1601

Banerjee AK (1980) 5'-terminal cap structure in eukaryotic messenger ribonucleic acids. Microbiol Rev 44: 175–205

Berleth T, Burri M, Thoma G, Bopp D, Richstein S, Frigerio G, Noll M, Nüsslein-Volhard C (1988) The role of localization of bicoid RNA in organizing the anterior pattern of the Drosophila embryo. EMBO J 7: 1749–1756

Bonneau A-M, Sonenberg N (1987) Involvement of the 24-kDa cap-binding protein in regulation of protein synthesis in mitosis. J Biol Chem 262: 11134–11139

Brenner C, Nakayama N, Goebi M, Tanaka K, Toh-e A, Matsumoto K (1988) CDC33 encodes mRNA cap-binding protein eIF-4E of Saccharomyces cerevisiae. Mol Cell Biol 8: 3556–3559

Bruenn J, Keitz B (1976) The 5' ends of yeast killer factor RNAs are pppGp. Nucleic Acids Res 3: 2427–2436

Bruenn JA (1980) Virus-like particles of yeast. Annu Rev Microbiol 34: 49–68

Canaani KH, Revel M, Groner Y (1976) Translational discrimination of 'capped' and 'noncapped' mRNAs: inhibition by a series of chemical analogs of m^7 GpppX. FEBS Lett 64: 326–331

Cavener DR (1987) Comparison of the consensus sequence flanking translational start sites in Drosophila and vertebrates. Nucleic Acids Res 15: 1353–1361

Cavener DR, Cavener BA (1993) Translation start sites and mRNA leaders. In: Maroni G (ed) An atlas of drosophila genes. Oxford University Press, Oxford, pp 359–377

Chakravarti D, Maitra U (1993) Eukaryotic translation initiation factor 5 from Saccharomyces cerevisiae. Cloning, characterization, and expression of the gene encoding the 45, 346-Da protein. J Biol Chem 268: 10524–10533

Chen C, Sarnow P (1995) Initiation of protein synthesis by the eukaryotic translational apparatus on circular RNAs. Science 268: 415–417

Chen C, Macejak DG, Oh S-K, Sarnow P (1993) Translation initiation by internal ribosome binding of eukaryotic mRNA molecules. In: Nierhaus KN, Franceschi F, Subramanian AR, Erdmann VA, Wittmann-Liebold B (eds) The translational apparatus. Plenum, New York, pp 229–240

Cigan AM, Donahue TF (1987) Sequence and structural features associated with translational initiator regions in yeast-a review. Gene 59: 1–18

Cigan AM, Feng L, Donahue TF (1988a) $tRNA_i^{met}$ functions in directing the scanning ribosome to the start site of translation. Science 242: 93–96

Cigan AM, Pabich EK Donahue TF (1988b) Mutational analysis of the His4 translational initiator region in Saccharomyces cerevisiae. Mol Cell Biol 8: 2964–2975

Cigan AM, Pabich EK, Feng L, Donahue TF (1989) Yeast translation initiation suppressor sui2 encodes the α subunit of eukaryotic initiation factor 2 and shares sequence identity with the human α subunit. Proc Natl Acad Sci USA 86: 2784–2788

Cigan AM, Bushman JL, Boal TR, Hinnebusch AG (1993) A protein complex of translation factor 2 in yeast. Proc Natl Acad Sci USA 90: 5350–5354

Coppolecchia R, Buser P, Stotz A, Linder P (1993) A new yeast translation initiation factor suppresses a mutation in the eIF-4A helicase. EMBO J 12: 4005–4011

Coward P, Dasgupta A (1992) Yeast cells are incapable of translating RNAs containing the poliovirus 5' untranslated region: evidence for a translational inhibitor. J Virol 66: 286–295

De Kloet SR, Andrean BAG (1976) Methylated nucleosides in polyadenylated-containing yeast messenger ribonucleic acid. Biochim Biophis Acta 425: 401–408

Dever TE, Feng L, Wek RC, Cigan AM, Donahue TF, Hinnebusch AG (1992) Phosphorylation of initiation factor 2 alpha by protein kinase GCN2 mediates gene-specific translational control of GCN4 in yeast. Cell 68: 585–596

Dolph PJ, Racaniello VR, Villamarin A, Palladino F, Schneider RJ (1988) The adenovirus tripartite leader eliminates the requirement for cap binding during translation initiation. J Virol 62: 2059–2066

Dolph PJ, Huang J, Schneider RJ (1990) Translation by the adenovirus tripartite leader: elements which determine independence from cap-binding complex. J Virol 64: 2669–2677

Donahue TF, Cigan AM, Pabich EK, Castilho-Valavicius B (1988) Mutations at a Zn(II) finger motif in the yeast eIF-2β gene alter ribosomal start-site selection during the scanning process. Cell 54: 621–632

Edery I, Sonenberg N (1985) Cap-dependent RNA splicing RNA splicing in HeLa nuclear extracts. Proc Natl Acad Sci USA 82: 7590–7594

Edery I, Humbelin M, Draveau D, Lee KAW, Milburn S, Hershey JWB, Trachsel H, Sonenberg N (1983) Involvement of eIF-4A in the cap recognition process. J Biol Chem 258: 11398–11403

Etchison D, Etchison JR (1987) Monoclonal antibody-aided characterization of cellular p220 in uninfected and poliovirus-infected HeLa cells: subcellular distribution and indentification of conformers. J Virol 61: 2702–2710

Etchison D, Milbourn SC, Edery I, Sonenberg N, Hershey JWB (1982) Inhibition of HeLa cell protein synthesis following poliovirus infection correlates with the proteolysis of a 220,000-dalton polypeptide associated with eukaryotic initiation factor 3 and cap binding protein complex. J Biol Chem 251: 14806–14810

Everett JG, Gallie DR (1992) RNA delivery in Saccharomyces cerevisiae using electroporation. Yeast 8: 1007–1014

Furuichi Y, LaFiandra A, Shatkin AJ (1977) 5'-Terminal structure and mRNA stability. Nature 266: 235–239

Fütterer J, Kiss-Laszlo Z, Hohn T (1993) Nonlinear ribosome migration on cauliflower mosaic virus 35S RNA. Cell 73: 789–802

Gallie DR (1991) The cap and poly(A) tail function synergistically to regulate mRNA translational efficiency. Genes Dev 5: 2108–2116

Georgiev O, Mous J, Birnstiel ML (1984) Processing and nucleo-cytoplasmic transport of histone gene transcripts. Nucleic Acids Res 12: 8539–8551

Goyer C, Altmann M, Lee HS, Blanc A, Deshmukh M, Woolford J, Trachsel H, Sonenberg N (1993) TIF4631 and TIF4632: two yeast genes encoding the high-molecular-weight subunits of the cap-binding protein complex (eukaryotic initiation factor 4F) contain an RNA recognition motif-like sequence and carry out an essential function Mol Cell Biol 13: 4860–4874

Green MR, Maniatis T, Melton DA (1983) Human β-globin pre-mRNA synthesized in vitro is accurately spliced in Xenopus oocyte. Cell 31: 681–694

Grifo JA, Tahara SM, Morgan MA, Shatkin AJ, Merrick WC (1983) New initiation factor activity required for globin mRNA translation. J Biol Chem 258: 5804–5810

Gulyas KD, Donahue TF (1992) SSL2, a suppressor of a stem-loop mutation in the HIS4 leader encodes the yeast homolog of human ERCC-3. Cell 69: 1031–1042

Hanic-Joyce P, Singer RA, Johnston GC (1987) Molecular characterization of the yeast PRT1 gene in which mutations affect translation initiation and regulation of cell proliferation. J Biol Chem 262: 2845–2851

Hannig EM, Cigan AM, Freeman BA, Kinzy TG (1993) GCD11, a negative regulator of GCN4 expression, encodes the γ-subunit of eIF-2 in Saccharomyces cerevisiae. Mol Cell Biol 13: 506–520

Hellen CUT, Witherell GW, Schmid M, Shin SH, Pestova TV, Gil A, Wimmer E (1993) A cytoplasmic 57kDa protein that is required for translation of picornavirus RNA by internal ribosomal entry is identical to the nuclear polypyrimidine tract-binding protein. Proc Natl Acad Sci USA 90: 7642–7646

Herman RC (1986) Internal initiation of translation on the vesicular stomatitis virus phosphoprotein mRNA yields a second protein. J Virol 58: 797–804

Herman RC (1987) Characterization of the internal initiation of translation on the vesicular stomatitis virus phosphoprotein mRNA. Biochemistry 26: 8346–8350

Hershey JWB (1991) Translational control in mammalian cells. Annu Rev Biochem 60: 717–755

Hinnebusch AG (1990) Transcriptional and translational regulation of gene expression in the general control of amino-acid biosynthesis in Saccharomyces cerevisiae. Prog Nucleic Acid Res Mol Biol 38: 195–240

Hinnebusch AG, Liebman SW (1991) Protein synthesis and translational control in Saccharomyces cerevisiae. In: Broach RJ, Pringle JR, Jones EW (eds) The molecular and cellular biology of the yeast Saccharomyces, vol 1. Cold Spring Harbor Laboratory Press, Cold Spring Harbor, pp 627–735

Hooper JE, Pérez-Alonso M, Bermingham JR, Prout M, Rocklein BA, Wagenbach M, Edstrom J-E, de Frutos R, Scott MP (1992) Comparative studies of drosophila Antennapedia genes. Genetics 132: 453–469

Hsu CL, Stevens A (1993) Yeast cells lacking 5'->3' exoribonuclease 1 contain mRNA species that are poly(A) deficient and partially lack the 5' cap structure. Mol Cell Biol 13: 4826–4835

Huang J, Schneider RJ (1991) Adenovirus inhibition of cellular protein synthesis involves inactivation of cap-binding protein. Cell 65: 271–280

Iizuka N, Najita L, Franzusoff A, Sarnow P (1994) Cap-dependent and cap-independent translation translational by internal initiation cell-free extracts prepared from Saccharomyces cerevisiae. Mol Cell Biol 14: 7322–7330

Jang SK, Wimmer E (1990) Cap-independent translation of encephalomyocarditis virus RNA: structural elements of the internal ribosomal entry site and involvement of a cellular 57-kD RNA-binding protein. Genes Dev 4: 1560–1572

Jang SK, Krausslich HG, Nicklin MJH, Duke GM, Palmenberg AC, Wimmer E (1988) A segment of the 5' nontranslated region of encephalomyocarditis virus RNA directs internal entry of ribosomes during in vitro translation. J Virol 62: 2636–2643

Kleppe K, van de Sande JH, Khorana HG (1970) Polynucleotide ligase-catalyzed joining of deoxyribo-oligo on ribopolynucleotide template. Proc Natl Acad Sci USA 67: 68–73

Konarska M, Filipowicz W, Domdey H, Gross H (1981) Binding of ribosomes to linear and circular forms of the 5'-terminal leader fragment of tobacco-mosaic-virus RNA. Eur J Biochem 114: 221–227

Konarska MM, Padgett RA, Sharp PA (1984) Recognition of cap structure in splicing in vitro of mRNA precursors. Cell 38: 731–736

Kozak M (1979) Inability of circular mRNA to attach to eukaryotic ribosomes. Nature 280: 82–85

Kozak M (1986) Point mutations define a sequence flanking the initiator codon that modulates translation by eukaryotic ribosomes. Cell 44: 283–292

Kozak M (1987a) An analysis of 5'-noncoding sequences upstream from 699 vertebrate messenger RNAs. Nucleic Acids Res 15: 8125–8148

Kozak M (1987b) At least six nucleotides preceding the AUG initiator codon enhance translation in mammalian cells. J Mol Biol 196: 947–950

Kozak M (1989) The scanning model for translation: an update. J Cell Biol 108: 229–241

Kozak M (1991) An analysis of vertebrate mRNA sequences: Intimations of translational control. Cell Biol 115: 887–903

Laughon A, Boulet AM, Bermingham JR, Laymon RA, Scott MP (1986) Structure of transcripts from the homeotic Antennapedia gene of Drosophila melanogaster two promoters control the major protein-coding region. Mol Cell Biol 6: 4676–4689

Lee AS, Delegeane A, Scharff D (1981) Highly conserved glucose-regulated protein in hamster and chicken cells: preliminary characterization of its cDNA clone. Proc Natl Acad Sci USA 78: 4922–4925

Lee AS, Bell J, Ting J (1984) Biochemical characterization of the 94- and 78-kilodalton glucose-regulated proteins in hamster fibroblasts. J Biol Chem 259: 4616–4621

Lejbkowicz F, Goyer C, Draveau A, Neron S, Lemieux R, Sonenberg N (1992) A fraction of the mRNA 5' cap-binding protein, eukaryotic initiation factor 4E, localizes to the nucleus. Proc Natl Acad Sci USA 89: 9612–9616

Linder P, Slonimski PP (1989) An essential yeast protein, encoded by duplicated genes TIF1 and TIF2 and homologous to the mammalian translation initiation factor eIF-4A, can suppress a mitochondrial missense mutation. Proc Natl Acad Sci USA 86: 2286–2290

Macejak DG, Sarnow P (1991) Internal initiation of translation mediated by the 5' leader of a cellular mRNA. Nature 353: 90–94

Macejak DG, Hambidge SJ, Najita LM, Sarnow P (1990). EIF-4F-independent translation of poliovirus RNA and cellular mRNA encoding glucose-regulated protein 78/immunoglobulin heavy-chain binding protein. Brinton MA, Heinz FX (eds) New aspects of positive-stranded RNA viruses. American Society for Microbiology, Washington DC, pp 152–157

Maicas E, Shago M, Friesen JD (1990) Translation of the Saccharomyces cerevisiae tcm 1 gene in the absence of a 5' untranslated leader. Nucleic Acids Res 18: 5823–5828

Maroto FG, Sierra JM (1988) Translational control in heat-shocked Drosophila embryos. J Biol Chem 263: 15720–15725

Marton MJ, Crouch K, Hinnebusch AG (1993) GCN1, a translational activatior of GCN4 in Saccharomyces cerevisiae, is required for phosphorylation of eukaryotic translation initiation factor 2 by protein kinase GCN2. Mol Cell Biol 13: 3541–3556

Meerovitch K, Svitkin YV, Lee HS, Lejbkowicz F, Kenan DJ, Chan EKL, Agol VI, Keene, JD, Sonenberg N (1993) La autoantigen enhances and corrects aberrant translation of poliuovirus RNA in Reticulocyte lysate. J Virol 67: 3798–3807

Merrick WC (1992) Mechanism and regulation of eukaryotic protein synthesis. Microbiol Rev 56: 291–315

Molla A, Jang SK, Paul AV, Reuer Q, Wimmer E (1992) Cardioviral internal ribosomal entry site is functional in a genetically engineered dicistronic poliovirus. Nature 356: 255–257

Molla A, Paul AV, Schmid M, Jang SK, Wimmer E (1993) Studies on dicistronic polioviruses implicate viral proteinase $2A^{pro}$ in RNA replication. Virol 196: 739–747

Moore MJ, Sharp PA (1992) Site-specific modification of pre-mRNA: the 2'-hydroxyl groups at the splice sites. Science 256: 992–997

Muhlrad D, Decker CJ, Parker R (1994) Deadenylation of the unstable mRNA encoded by the yeast MFA2 gene leads to decapping followed by 5'3' digestion of the transcript. Genes Dev 8: 855–866

Munoz A, Alonso MA, Carrasco L (1984) Synthesis of heat shock proteins in HeLa cells: inhibition by virus infection. Virology 137: 150–159

Munro S, Pelham HRB (1986) An hsp 70-like protein in the ER: Identity with the 78kD glucose-regulated protein and immunoglobulin heavy chain binding protein. Cell 46: 291–300
Nomoto A, Lee YF, Wimmer E (1976) The 5' end of poliovirus mRNA is not capped with m^{7}G(5')ppp(5')Np. Proc Natl Acad Sci USA 73: 375–380
Nomoto A, Kitamura N, Golini F, Wimmer E (1977) The 5'-terminal structures of poliovirion RNA and poliovirus mRNA differ only in the genome-linked protein. VPg. Proc Natl Acad Sci USA 74: 5345–5349
Oh SK, Scott MP, Sarnow P (1992) Homeotic gene antennapedia mRNA contains 5'-noncoding sequences that confer translational initiation by internal ribosome binding. Genes Dev 6: 1643–1653
Pelletier J, Sonenberg N (1988) Internal initiation of translation of eukaryotic mRNA directed by a sequence derived from poliovirus RNA. Nature 334: 320–325
Pelletier J, Sonenberg N (1989) Internal binding of eukaryotic ribosomes on poliovirus RNA: translation in Hela cell extracts. J Virol 63: 441–444
Romaniuk PJ, Uhlenbeck OC (1983) Joining of RNA molecules with RNA ligase. Methods Enzymol 100: 52–59
Rozen F, Edery I, Meerovitch K, Dever TE, Merrcik WC, Sonenberg N (1990) Bidirectional RNA helicase activity of eukaryotic initiation factor 4A and 4F. Mol Cell Biol 10:1134–1144
Russell PJ, Hambidge SJ, Kirkegaard K (1991) Direct introduction and transient expression of capped and non-capped RNA is Saccharomyces Cerevisiae. Nucleic Acids Res 19: 4949–4953
Sachs AB, Davis RW (1989) The poly(A) binding protein is required for poly(A) shortening and 60S ribosomal subunit-dependent translation initiation. Cell 58: 857–867
Sachs AB, Deardorff JA (1992) Translation initiation requires the PAB-dependent poly(A) ribonuclease in yeast. Cell 70: 961–973
Sankar S, Cheah K-C, Porter AG (1989) Antisense oligonucleotide inhibition of encephalomyocarditis virus RNA translation. Eur J Biochem 184: 465–480
Sarnow P (1989) Translation of glucose-regulated protein 78/immunoglobulin heavy-chain binding protein mRNA is increased in poliovirus-infected cells at a time when cap-dependent translation of cellular mRNAs inhibited. Proc Natl Acad Sci USA 86: 5795–5799
Schnier J, Schwelberger HG, Smit-McBride Z, Kang HA, Hershey WB (1991) Translation initiation factor 5A and its hypusine modification are essential for cell viability in the yeast Saccharomyces cerevisiae. Mol Cell Biol 11: 3105–3114
Shatkin AJ (1976) Capping of eukaryotic mRNAs. Cell 9: 645–653
Sherman F, Stewart JW (1982) Mutations altering initiation of translation of yeast iso-1-cytochrome c; contrasts between the eukaryotic and prokaryotic initiation process. In: Strathern JN, Jones EW, Broach JR (eds) The molecular biology of the yeast Saccharomyces: metabolism and gene expression. Cold Spring Harbor Laboratory Press, Cold Spring Harbor, pp 301–333
Shih DS, Park IW, Evans CL, Jaynes JM, Palmenberg AC (1987) Effects of cDNA hybridization on translation of encephalomyocarditis virus RNA. J Virol 61: 2033–2037
Sripati CE, Groner Y, Warner JR (1976) Methylated, blocked 5' termini of yeast mRNA. J Biol Chem 251: 2898–2904
Stevens A (1978) An exoribonuclease from Saccharomyces cerevisiae: effect of modifications of 5' end groups on the hydrolysis of substrates to 5'-nucleotides. Biochem Biophys Res Commun 81: 656–661
Stevens A, Maupin MK (1987) A 5' to 3' exoribonuclease of Saccharomyces cerevisiae: Size and novel substrate specificity. Arch Biochem Biophys 252: 339–347
Stroeher VL, Jorgensen EM, Garber RL (1986) Multiple transcripts from the Antennapedia gene of Drosophila. Mol Cell Biol 6: 4667–4675
Tahara SM, Morgan MA, Shatkin AJ (1981) Two forms of purified m7G cap-binding protein with different effects on capped mRNA translation in extracts of uninfected and poliovirus-infected HeLa cells. J Biol Chem 256: 7691–7694
Ting J, Lee AS (1988) Human gene encoding the 78,000-dalton glucose-regulated protein and its pseudogene: structure, conservation, and regulation. DNA 7: 275–286
Tsukiyama-Kohara K, Iizuka N, Kohara M, Nomoto A (1992) Internal ribosome entry site within hepatitis C virus RNA. J Virol 66: 1476–1483
Van den Heuvel JJ, Bergkamp RJM, Planta RJ Raué HA (1989) Effect of deletions in the 5'-noncoding region on the translational efficiency of phosphoglycerate kinase mRNA in yeast. Gene 79: 83–95

Vega Laso MR, Zhu D, Sagliocco F, Brown AP, Tuite MF, McCarthy JEG (1993) Inhibition of translational initiation in the yeast Saccharomyces cerevisiae as a function of the stability and position of hairpin structures in the mRNA leader. J Biol Chem 268: 6453–6462
Wang C, Lehmann R (1991) Nanos is the localized posterior determinant in Drosophila. Cell 66: 637–647
Wang C, Sarnow P, Siddiqui A (1993) Translation of human hepatitis C virus RNA in cultured cells is mediated by an internal ribosome-binding mechanism. J Virol 67: 3338–3344
Weber LA, Hickey ED, Baglioni C (1987) Influence of potassium salt concentration and temperature on inhibition of translation by 7-methylguanosine-5'-monophosphate. J Biol Chem 253: 178–183
Wickner RB (1992) Double-stranded and single-stranded RNA viruses of Saccharomyces cerevisiae. Annu Rev Microbiol 46: 347–375
Widner WR, Wickner RB (1993) Evidence that the SKI antiviral system of Saccharomyces cerevisiae acts by blocking expression of viral mRNA. Mol Cell Biol 13: 4331–4341
Yoon H, Donahue TF (1992) The sui1 suppressor locus in Saccharomyces cerevisiae encodes a translation factor that functions during $tRNA_i^{Met}$ recognition of the start codon. Mol Cell Biol 12: 248–260
Yoon H, Miller SP, Pabich EK, Donahue TF (1992) SSL1, a suppressor of a HIS4 5'UTR stem-loop mutation is, essential for translation initiation and affects UV resistance in yeast. Genes Dev 6: 2463–2477
Zhong T, Arndt KT (1993) The yeast SIS1 protein, a DnaJ homolog, is required for the initiation of translation. Cell 73: 1175–1186
Zuker M, Stiegler P (1981) Optimal computer folding of large RNA sequences using thermodynamics and auxiliary information. Nucleic Acids Res 9: 133–148

Subject Index

Current Topics in Microbiology and Immunology

Volumes published since 1989 (and still available)

Vol. 160: **Oldstone, Michael B. A.; Koprowski, Hilary (Eds.):** Retrovirus Infections of the Nervous System. 1990. 16 figs. XII, 176 pp. ISBN 3-540-51939-4

Vol. 161: **Racaniello, Vincent R. (Ed.):** Picornaviruses. 1990. 12 figs. X, 194 pp. ISBN 3-540-52429-0

Vol. 162: **Roy, Polly; Gorman, Barry M. (Eds.):** Bluetongue Viruses. 1990. 37 figs. X, 200 pp. ISBN 3-540-51922-X

Vol. 163: **Turner, Peter C.; Moyer, Richard W. (Eds.):** Poxviruses. 1990. 23 figs. X, 210 pp. ISBN 3-540-52430-4

Vol. 164: **Bækkeskov, Steinnun; Hansen, Bruno (Eds.):** Human Diabetes. 1990. 9 figs. X, 198 pp. ISBN 3-540-52652-8

Vol. 165: **Bothwell, Mark (Ed.):** Neuronal Growth Factors. 1991. 14 figs. IX, 173 pp. ISBN 3-540-52654-4

Vol. 166: **Potter, Michael; Melchers, Fritz (Eds.):** Mechanisms in B-Cell Neoplasia. 1990. 143 figs. XIX, 380 pp. ISBN 3-540-52886-5

Vol. 167: **Kaufmann, Stefan H. E. (Ed.):** Heat Shock Proteins and Immune Response. 1991. 18 figs. IX, 214 pp. ISBN 3-540-52857-1

Vol. 168: **Mason, William S.; Seeger, Christoph (Eds.):** Hepadnaviruses. Molecular Biology and Pathogenesis. 1991. 21 figs. X, 206 pp. ISBN 3-540-53060-6

Vol. 169: **Kolakofsky, Daniel (Ed.):** Bunyaviridae. 1991. 34 figs. X, 256 pp. ISBN 3-540-53061-4

Vol. 170: **Compans, Richard W. (Ed.):** Protein Traffic in Eukaryotic Cells. Selected Reviews. 1991. 14 figs. X, 186 pp. ISBN 3-540-53631-0

Vol. 171: **Kung, Hsing-Jien; Vogt, Peter K. (Eds.):** Retroviral Insertion and Oncogene Activation. 1991. 18 figs. X, 179 pp. ISBN 3-540-53857-7

Vol. 172: **Chesebro, Bruce W. (Ed.):** Transmissible Spongiform Encephalopathies. 1991. 48 figs. X, 288 pp. ISBN 3-540-53883-6

Vol. 173: **Pfeffer, Klaus; Heeg, Klaus; Wagner, Hermann; Riethmüller, Gert (Eds.):** Function and Specificity of γ/δ TCells. 1991. 41 figs. XII, 296 pp. ISBN 3-540-53781-3

Vol. 174: **Fleischer, Bernhard; Sjögren, Hans Olov (Eds.):** Superantigens. 1991. 13 figs. IX, 137 pp. ISBN 3-540-54205-1

Vol. 175: **Aktories, Klaus (Ed.):** ADP-Ribosylating Toxins. 1992. 23 figs. IX, 148 pp. ISBN 3-540-54598-0

Vol. 176: **Holland, John J. (Ed.):** Genetic Diversity of RNA Viruses. 1992. 34 figs. IX, 226 pp. ISBN 3-540-54652-9

Vol. 177: **Müller-Sieburg, Christa; Torok-Storb, Beverly; Visser, Jan; Storb, Rainer (Eds.):** Hematopoietic Stem Cells. 1992. 18 figs. XIII, 143 pp. ISBN 3-540-54531-X

Vol. 178: **Parker, Charles J. (Ed.):** Membrane Defenses Against Attack by Complement and Perforins. 1992. 26 figs. VIII, 188 pp. ISBN 3-540-54653-7

Vol. 179: **Rouse, Barry T. (Ed.):** Herpes Simplex Virus. 1992. 9 figs. X, 180 pp. ISBN 3-540-55066-6

Vol. 180: **Sansonetti, P. J. (Ed.):** Pathogenesis of Shigellosis. 1992. 15 figs. X, 143 pp. ISBN 3-540-55058-5

Vol. 181: **Russell, Stephen W.; Gordon, Siamon (Eds.):** Macrophage Biology and Activation. 1992. 42 figs. IX, 299 pp. ISBN 3-540-55293-6

Vol. 182: **Potter, Michael; Melchers, Fritz (Eds.):** Mechanisms in B-Cell Neoplasia. 1992. 188 figs. XX, 499 pp. ISBN 3-540-55658-3

Vol. 183: **Dimmock, Nigel J.:** Neutralization of Animal Viruses. 1993. 10 figs. VII, 149 pp. ISBN 3-540-56030-0

Vol. 184: **Dunon, Dominique; Mackay, Charles R.; Imhof, Beat A. (Eds.):** Adhesion in Leukocyte Homing and Differentiation. 1993. 37 figs. IX, 260 pp. ISBN 3-540-56756-9

Vol. 185: **Ramig, Robert F. (Ed.):** Rotaviruses. 1994. 37 figs. X, 380 pp. ISBN 3-540-56761-5

Vol. 186: **zur Hausen, Harald (Ed.):** Human Pathogenic Papillomaviruses. 1994. 37 figs. XIII, 274 pp. ISBN 3-540-57193-0

Vol. 187: **Rupprecht, Charles E.; Dietzschold, Bernhard; Koprowski, Hilary (Eds.):** Lyssaviruses. 1994. 50 figs. IX, 352 pp. ISBN 3-540-57194-9

Vol. 188: **Letvin, Norman L.; Desrosiers, Ronald C. (Eds.):** Simian Immunodeficiency Virus. 1994. 37 figs. X, 240 pp. ISBN 3-540-57274-0

Vol. 189: **Oldstone, Michael B. A. (Ed.):** Cytotoxic T-Lymphocytes in Human Viral and Malaria Infections. 1994. 37 figs. IX, 210 pp. ISBN 3-540-57259-7

Vol. 190: **Koprowski, Hilary; Lipkin, W. Ian (Eds.):** Borna Disease. 1995. 33 figs. IX, 134 pp. ISBN 3-540-57388-7

Vol. 191: **ter Meulen, Volker; Billeter, Martin A. (Eds.):** Measles Virus. 1995. 23 figs. IX, 196 pp. ISBN 3-540-57389-5

Vol. 192: **Dangl, Jeffrey L. (Ed.):** Bacterial Pathogenesis of Plants and Animals. 1994. 41 figs. IX, 343 pp. ISBN 3-540-57391-7

Vol. 193: **Chen, Irvin S. Y.; Koprowski, Hilary; Srinivasan, Alagarsamy; Vogt, Peter K. (Eds.):** Transacting Functions of Human Retroviruses. 1995. 49 figs. IX, 240 pp. ISBN 3-540-57901-X

Vol. 194: **Potter, Michael; Melchers, Fritz (Eds.):** Mechanisms in B-cell Neoplasia. 1995. 152 figs. XXV, 458 pp. ISBN 3-540-58447-1

Vol. 195: **Montecucco, Cesare (Ed.):** Clostridial Neurotoxins. 1995. 28 figs. XI., 278 pp. ISBN 3-540-58452-8

Vol. 196: **Koprowski, Hilary; Maeda, Hiroshi (Eds.):** The Role of Nitric Oxide in Physiology and Pathophysiology. 1995. 21 figs. IX, 90 pp. ISBN 3-540-58214-2

Vol. 197: **Meyer, Peter (Ed.):** Gene Silencing in Higher Plants and Related Phenomena in Other Eukaryotes. 1995. 17 figs. IX, 232 pp. ISBN 3-540-58236-3

Vol. 198: **Griffiths, Gillian M.; Tschopp, Jürg (Eds.):** Pathways for Cytolysis. 1995. 45 figs. IX, 224 pp. ISBN 3-540-58725-X

Vol. 199/I: **Doerfler, Walter; Böhm, Petra (Eds.):** The Molecular Repertoire of Adenoviruses I. 1995. 51 figs. XIII, 280 pp. ISBN 3-540-58828-0

Vol. 199/II: **Doerfler, Walter; Böhm, Petra (Eds.):** The Molecular Repertoire of Adenoviruses II. 1995. 36 figs. XIII, 278 pp. ISBN 3-540-58829-9

Vol. 199/III: **Doerfler, Walter; Böhm, Petra (Eds.):** The Molecular Repertoire of Adenoviruses III. 1995. 50 figs. Approx. XIII, 324 pp. ISBN 3-540-58987-2

Vol. 200: **Kroemer, Guido; Martinez-A., Carlos (Eds.):** Apoptosis in Immunology. 1995. 14 figs. XI, 242 pp. ISBN 3-540-58756-X

Vol. 201: **Kosco-Vilbois, Marie H. (Ed.):** An Antigen Depository of the Immune System: Follicular Dendritic Cells. 1995. 39 figs. IX, 209 pp. ISBN 3-540-59013-7

Vol. 202: **Oldstone, Michael B. A.; Vitkovic, Ljubisa (Eds.):** HIV and Dementia. 1995. 40 figs. XIII, 283 pp. ISBN 3-540-59117-6

Springer-Verlag and the Environment

MIX
Papier aus verantwortungsvollen Quellen
Paper from responsible sources
FSC® C105338

If you have any concerns about our products,
you can contact us on
ProductSafety@springernature.com

In case Publisher is established outside the EU,
the EU authorized representative is:
Springer Nature Customer Service Center GmbH
Europaplatz 3, 69115 Heidelberg, Germany

Printed by Libri Plureos GmbH
in Hamburg, Germany